AF775233

Beyond Rain - The Future of Irrigation in a Changing Climate

Edited by Daniel Bucur

Published in London, United Kingdom

Beyond Rain - The Future of Irrigation in a Changing Climate
http://dx.doi.org/10.5772/intechopen.1005725
Edited by Daniel Bucur

Contributors
Alem Oyarmoi, Stephen Birkinshaw, Caspar J. M. Hewett, Hayley J. Fowler, Vallal Kannan Sankaralingam, Marimuthu Subramanian, Sangeetha Jebalin Vijayalalitha Vallal Kannan, HariniHarishma Veerappan Prithviraj, Agwu A. Ejem, Tarek Mahrous, Ali Mohammed Abd-Elkader, Ertuğrul Karaş

© The Editor(s) and the Author(s) 2025
The rights of the editor(s) and the author(s) have been asserted in accordance with the Copyright, Designs and Patents Act 1988. All rights to the book as a whole are reserved by INTECHOPEN LIMITED. The book as a whole (compilation) cannot be reproduced, distributed or used for commercial or non-commercial purposes without INTECHOPEN LIMITED's written permission. Enquiries concerning the use of the book should be directed to INTECHOPEN LIMITED rights and permissions department (permissions@intechopen.com).
Violations are liable to prosecution under the governing Copyright Law.

Individual chapters of this publication are distributed under the terms of the Creative Commons Attribution 4.0 License which permits commercial use, distribution and reproduction of the individual chapters, provided the original author(s) and source publication are appropriately acknowledged. If so indicated, certain images may not be included under the Creative Commons license. In such cases users will need to obtain permission from the license holder to reproduce the material. More details and guidelines concerning content reuse and adaptation can be found at http://www.intechopen.com/copyright-policy.html.

Notice
Statements and opinions expressed in the chapters are these of the individual contributors and not necessarily those of the editors or publisher. No responsibility is accepted for the accuracy of information contained in the published chapters. The publisher assumes no responsibility for any damage or injury to persons or property arising out of the use of any materials, instructions, methods or ideas contained in the book.

First published in London, United Kingdom, 2025 by IntechOpen
IntechOpen is the global imprint of INTECHOPEN LIMITED, registered in England and Wales, registration number: 11086078, 167-169 Great Portland Street, London, W1W 5PF, United Kingdom

For EU product safety concerns: IN TECH d.o.o., Prolaz Marije Krucifikse Kozulić 3, 51000 Rijeka, Croatia, info@intechopen.com or visit our website at intechopen.com.

British Library Cataloguing-in-Publication Data
A catalogue record for this book is available from the British Library

Beyond Rain - The Future of Irrigation in a Changing Climate
Edited by Daniel Bucur
p. cm.

This title is part of the Agricultural Sciences Book Series, Volume 27
Topic: Agronomy and Horticulture
Series Editor: W. James Grichar
Topic Editor: Ibrahim Kahramanoglu
Associate Topic Editors: Murat Helvaci and Olga Panfilova

Print ISBN 978-1-83634-578-7
Online ISBN 978-1-83634-577-0
eBook (PDF) ISBN 978-1-83634-579-4
ISSN 3029-052X

If disposing of this product, please recycle the paper responsibly.

IntechOpen
intechopen.com

Built by scientists, for scientists

Explore all IntechOpen books

IntechOpen Book Series
Agricultural Sciences
Volume 27

Aims and Scope of the Series

The importance of agriculture cannot be overstated. It helps sustain life, as it gives us the food we need to survive and provides opportunities for economic well-being. Agriculture helps people prosper around the world and combines the creativity, imagination, and skill involved in planting crops and raising animals with modern production methods and new technologies. This series includes two main topics: Agronomy and Horticulture, and Animal Farming. This series will help readers better understand the intricacies of production agriculture and provide the new knowledge that is required to be successful. The success of a farmer in modern agriculture requires knowledge of events happening locally as well as globally that impact input decisions and ultimately determine net profit.

Meet the Series Editor

W. James Grichar has been employed with Texas A&M AgriLife Research for over 45 years with an emphasis on research in agronomy, plant pathology, and weed science. He obtained his BS from Texas A&M in 1972 and his Masters of Plant Protection in 1975. He has published 195 journal articles, over 330 research reports and briefs, 11 book chapters, and over 300 abstracts of profession meetings. He also directs research in many crops including corn, grain sorghum, peanuts, and sesame. He has held various positions in different professional societies including the American Peanut Research and Education Society, Southern Weed Science Society, and Texas Plant Protection Conference in addition to being Associate Editor for Peanut Science and Weed Technology. Significant accomplishments have included spearheading efforts to determine the optimum planting time for soybean production along the upper Texas Gulf Coast. These efforts have shown growers that soybean yields can be improved by 10 to 20% by following a late March to early April plant date. He also has been instrumental in developing a herbicide program for peanut production in the south Texas growing region. Through the development and use of herbicides that are effective against major weed problems in the south Texas region, peanut yields have increased by 25 to 30%.

Meet the Volume Editor

Daniel Bucur is a Full Professor and Ph.D. supervisor at the 'Ion Ionescu de la Brad' Iasi University of Life Sciences (Romania), where he leads the Department of Pedotechnics within the Faculty of Agriculture. He has over 30 years of experience in higher education and research, focusing on soil management and sustainable agriculture. Dr. Bucur graduated in Land Reclamation in 1988 and earned his Ph.D. in 1994 from Gheorghe Asachi Technical University of Iasi. He is the author of 54 ISI-indexed articles, 8 books, and 6 international chapters, and serves as Editor-in-Chief of the *Journal of Applied Life Sciences and Environment*. He is also the editor of four international volumes published by IntechOpen. He is a corresponding member of the Romanian Academy of Agricultural and Forestry Sciences and an active member of the International Soil Tillage Research Organization (ISTRO), the International Union of Soil Sciences (IUSS), and the European Society for Soil Conservation (ESSC).

Contents

Preface

In a global context marked by accelerated climate change, demographic pressures, and increasingly limited natural resources, there must be profound and sustainable transformations to agriculture. Efficient water management has become a crucial condition for ensuring food security and conserving natural ecosystems, particularly in regions vulnerable to climatic extremes.

This volume brings together scientific and applied contributions that reflect the complexity of these challenges, addressing irrigation and productivity in agriculture from diverse and complementary perspectives. It examines the relationships between lowland irrigation practices and the hydrological balance of sensitive wetland ecosystems, as well as the broader implications of climate variability. The insights presented in this book offer valuable tools for policymakers and ecosystem managers to inform their adaptation of conservation strategies.

The discussion on practical solutions involves a comprehensive examination of modern irrigation technologies. These systems are accessible, sustainable, and capable of optimising water use while increasing crop yields. Among them, rain hose irrigation systems stand out for their efficiency, affordability, and ease of use, offering a viable option for small-scale farmers with limited access to advanced resources.

The social dimension of sustainable agriculture is explored through an integrated communication approach. The aim is to disseminate best practices, particularly within rural communities in less developed regions. There is an emphasis on tailoring communication methods and channels to local cultural and linguistic contexts to ensure inclusive adoption of sustainable techniques.

Next, this book turns towards a forward-looking perspective, showcasing soilless cultivation techniques – such as hydroponics, aeroponics, and aquaponics – as innovative solutions for resilient, urban, and resource-efficient agriculture. These technologies are empowered by digital tools, including artificial intelligence and the Internet of Things, which contribute to the emergence of circular and climate-smart agricultural systems.

Finally, this volume focuses on regional realities. It presents an applied analysis of soil and water management challenges in arid and semi-arid climates, with a particular focus on the Mediterranean zone. This discussion offers adaptable strategies for addressing water scarcity and extreme weather events through integrated and sustainable land-use practices.

As the editor of this volume, I would like to thank all of the contributing authors for their valuable work and the publisher for supporting this endeavour. The chapters that follow open up fertile ground for reflection and action, where science, technology, and human vision come together to shape the future of sustainable agriculture.

Daniel Bucur
Faculty of Agriculture,
Department of Pedotechnics,
'Ion Ionescu de la Brad' Iasi University of Life Sciences (IULS),
Iasi, Romania

Chapter 1

Effects of Lowland Rice Irrigation on Papyri Wetland Flow Regime under a Future Climate

Alem Oyarmoi, Stephen Birkinshaw, Caspar J.M. Hewett and Hayley J. Fowler

Abstract

Lowland rice is often cultivated in catchments dominated by wetlands. However, no study in the public domain has assessed the impacts of lowland rice irrigation on the most dominant wetland plant species in Africa (papyrus) under a future climate. Hydrological modelling results, complemented with ecologically relevant statistical analyses, indicate that climate change, due to the projected increase in precipitation over the East African region, will significantly influence wetland inflow more than irrigation for the irrigation command area of 2300 ha, equivalent to the irrigated area to catchment area ratio of the order 1:70. Like studies in other parts of the world, extreme flows are most affected, which could affect papyri reproduction. The extremes are particularly affected in frequency and duration. The 30-year median annual high and low pulse frequency is projected to drop from 15 to 18 and 17–22 to 1–3. However, the median yearly duration is projected to increase from 2 and 2–3 days to 35–45 days and 12–24 days. Overall, the findings in this study can help catchment managers in similar ecosystems plan for conservation programmes that can enhance the productivity of papyri, given that it is a keystone species.

Keywords: papyri wetlands, lowland rice, global warming, irrigation, high flows, low flows

1. Introduction

The World Economic Forum's 2023 Global Risks Report [1] ranked climate change and related environmental challenges as one of the top five global risks. Climate change affects marine, freshwater, terrestrial ecosystems, and ecosystem services [2]. Given global greenhouse gas emission rates, climate change will likely become the dominant direct driver of global biodiversity and ecosystem services loss by the end of the century [3]. As one of the most productive ecosystems, wetlands are particularly vulnerable to climate change [4], and the effects of climate change on wetland processes and fluxes are considered one of the most important unanswered questions for the foreseeable future [4, 5].

Wetland ecosystems maintain a delicate biogeochemical balance between soil, water, microbial, plant, and atmospheric processes [6]. Climate change modifies wetland biogeochemical processes through temperature increases and alteration of the hydrological regime [4, 7]. This affects essential wetland ecosystem services such as carbon sequestration, flow regulation, biodiversity conservation, and food supply. These climate-induced ecosystem changes are compounded by human activities such as river regulation, water abstraction, and land use change [8]. These human activities, particularly water abstraction and land use change, are mostly linked to agriculture [9–11] and are predicted to increase due to population growth and global warming [12–14].

The impacts of increased food demand and agricultural expansion on natural resources, including wetlands, could be minimised by adopting intensive agriculture [15]. Intensive agriculture focuses on increasing productivity on rainfall-constrained cropland through irrigation [16]. However, irrigation projects often threaten habitats and ecosystems, especially if developed without robust environment management plans [17], which climate change could further exacerbate by significantly altering both water availability and crop water requirement [18].

Studies show that Africa, with a growing population, has the potential to meet its future food demand through irrigation [12, 15], and several countries have placed irrigated agriculture on the development agenda [19, 20]. However, some countries have essential wetlands and habitats [21], which could be negatively impacted by climate change and irrigation water abstraction. Thus, as one of the most critical challenges scientists face today [4], there is a need to predict the impacts of future catchment changes on wetland habitats. However, it is challenging to predict the ecological consequences of coming change, given the complexity of the variables involved (e.g. environmental warming, alterations to hydrology, and biogeochemistry) [22, 23] and ecological data limitation (e.g. the composition, phenology, abundance, productivity, and distribution of these habitats under a future climate) [22].

Hydrological modelling helps predict how flows change due to irrigation or climate change [24, 25]. However, metrics that link hydrologic regimes (i.e. the magnitude, frequency, duration, timing, and rate of change of flows) of the changed flows to biologically relevant hydrologic attributes of ecosystems are required [26–29]. The indicators of hydrologic alteration (IHA) is a statistical approach for deriving biologically relevant hydrologic attributes [26, 30] and is the most widely used metric for the characterisation of ecologically relevant features of flow regimes [31]. The IHA uses the range of variability approach (RVA) in designing 'environmentally acceptable' flow regimes [32], providing a framework for setting preliminary flow management or restoration targets for rivers where annual streamflow characteristics frequently fall outside their historical range(s) of variability [32, 33].

As a low-cost approach to environmental flow management [34], the IHA-RVA concept has been applied globally to assess flow regime alterations due to agricultural water abstraction [35–39]. However, few studies, if any, have assessed the effects of lowland rice irrigation on wetlands under a future climate. This chapter assesses the impact of future climate and rice irrigation on papyrus-dominated wetlands in the Naigombwa catchment in Uganda. As Uganda's central rice food basket [40], the wetlands in this region of Uganda continually face the challenges of agricultural encroachment, mainly due to population pressure [41] and the effects of climate change [42]. For example, recently, to mitigate the impacts of climate change and offset the national rice food deficit, the Ministry of Agriculture, Animal Industry, and Fisheries established the Naigombwa rice irrigation project [43–45]. The scheme's construction works began in 2023 and are slated for completion by 2026. This study, therefore, has two main objectives: (1) to quantify the impacts of lowland rice irrigation and future climate on the natural inflow into papyrus-dominated wetlands; and (2) to understand how these changes to the flow regime will affect the sustainability of the wetlands.

2. Materials and methods

2.1 Study area

The Naigombwa catchment (33.41° to 33.76° East and 0.38° to 1.05° North) is a part of the wetland-dominated Mpologoma River basin [46] and covers approximately 1680 km^2. The Naigombwa wetland system is primarily located in the lower parts of the Naigombwa catchment (**Figure 1**) and covers roughly 31 km^2 (2%) of the catchment area. Typical of wetlands in the region, the papyrus-dominated marshes occur in flat, channel, and basin landforms (e.g. floodplains, rivers/creeks, and lakes, respectively [47]), with hydroperiods characterised by seasonal inundation (such as in floodplains) and permanent inundation (such as in rivers and lakes) [48].

The wetlands are habitats for several species of flora and fauna [49]. Non-papyrus plants, mainly vines that depend on the tall structural complexity of live papyrus [49], tend to occur as pockets of monoculture stands [48]. Animals, on the other hand, occupy both the lower water-saturated area (e.g. fish, worms, and insect larvae) and the upper surfaces (e.g. birds and insects) [48].

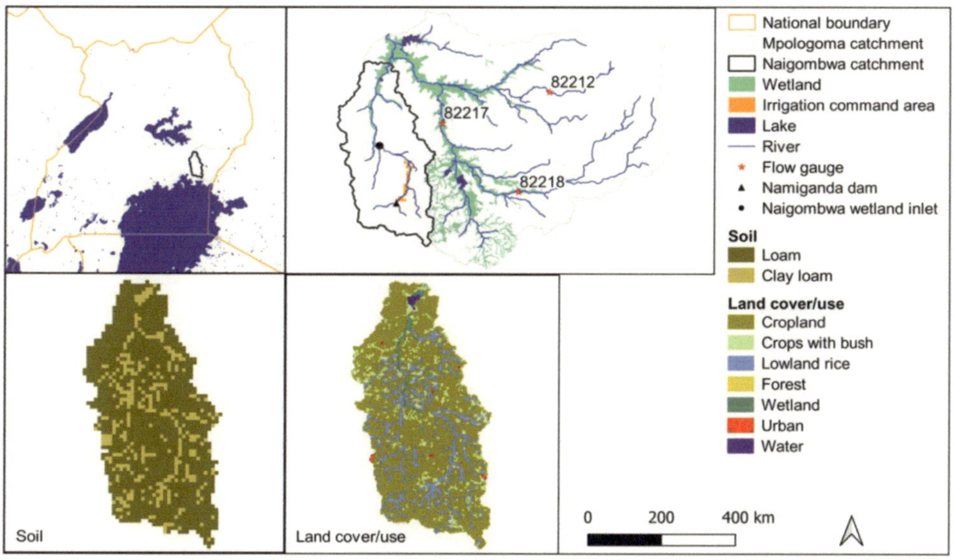

Figure 1.
Naigombwa catchment soils, land cover, irrigation command area, and location of the irrigation reservoir.

Phase 1 of the Naigombwa rice irrigation project will cover about 2300 hectares, including 2200 hectares of lowland rice and 100 hectares of polycrops (maize, beans, watermelon, etc.) [45]. The irrigation reservoir, Namiganda Dam, at 22.9 million m^3, will be located upstream of the irrigation command area.

2.2 Hydrological modelling

Since the irrigation command area is located upstream of the Naigombwa wetland (see **Figure 1**), the most effective way to assess the impact of the irrigation scheme on the wetland's flow regime is to analyse the inflow into the wetland. However, it is essential to note that the Naigombwa River does not have a gauge. As a result, the inflow into the wetlands was simulated using the SHETRAN catchment modelling tool. SHETRAN [50–52] is a 3D-integrated surface and subsurface finite difference modelling system for water flow, sediment transport, and contaminant transport in catchments. The model can simulate the impacts of various land covers, such as wetlands [53], and different water management practices such as irrigation and reservoirs [54]. The water flow components include interception and evapotranspiration, overland and channel flow, varyingly saturated lithology and aquifers, and channel-aquifer interactions (**Figure 2**). Detailed information on SHETRAN can be accessed at https://research.ncl.ac.uk/shetran/.

The SHETRAN catchment model by Oyarmoi et al. [53] describes the hydrology of the Mpologoma River basin (including the Naigombwa catchment) for current conditions (i.e. without irrigation and the Namiganda dam). This study modified the model to capture the influence of irrigation and the dam. The dam crest was set in the model to depict its design height of 11 m and spillway width and depth of 46.8 and 4 m, respectively [45]. The water used for irrigation was abstracted from the reservoir based on a predetermined irrigation schedule and applied to the irrigation grids in SHETRAN. Irrigation schedules were prepared for the traditional water-intensive continuous flooding (CF) [56, 57] and the less-intensive alternate wetting and drying (AWD) [58].

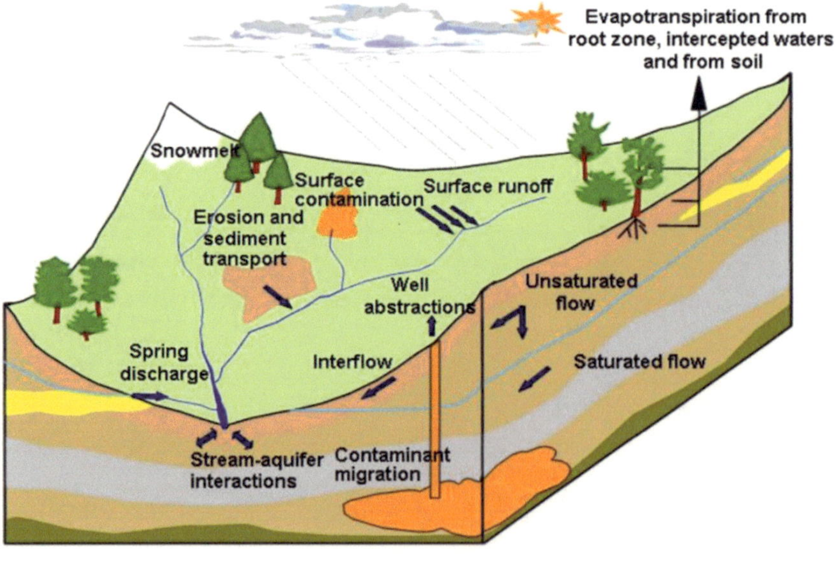

Figure 2.
The SHETRAN model catchment processes [55].

Lowland rice is typically cultivated in fields with bunds that retain ponded water. This was mimicked in SHETRAN by lowering the overland Strickler coefficient (the inverse of Manning's coefficient) when the irrigated grid water level is less than or equal to the bund height of 25 cm. The Strickler coefficient was lowered from the calibration value of 0.0469 $m^{1/3}/s$ for rice fields [53] to the minimal value in SHETRAN of 0.0001 $m^{1/3}/s$. As a result, overland flow ceases, but infiltration continues.

Hydrological simulations for the current and future climate were carried out using precipitation and potential evapotranspiration data from the Coupled Model Intercomparison Project phase 6 (CMIP6) Global Climate Models (GCMs). Model runs were executed with and without irrigation for the baseline (BL) and the 2 and 4°C global warming levels (i.e. GWL2 and GWL4), **Table 1**. The selection and bias correction of the CMIP6 GCMs used in this study (i.e. GFDL-ESM4, MRI-ESM2–0, and NorESM2-MM) is described in Oyarmoi et al. [53]. The SHETRAN model with irrigation allows quantifying the combined effects of irrigation and climate change at GWL2 and GWL4. Thus, for a particular warming level, the difference between the models with and without irrigation predicts the impact of irrigation alone on the flow regime.

2.2.1 Irrigation schedule and demand

There are two rain seasons in a year in the study area. The first rainy season is from March to May, and the second is from August to November. The lifecycle of annual crops such as rice, maize, beans, and watermelon are targeted to these

S. No.	Climatic datasets	Simulation period (model warmup period)	Irrigation present or not
1	GFDL-ESM4 at BL	1979–2013 (1979–1983)	Yes
2			No
3	MRI-ESM2-0 at BL	1979–2013 (1979–1983)	Yes
4			No
5	NorESM2-MM at BL	1979–2013 (1979–1983)	Yes
6			No
7	GFDL-ESM4 at GWL2	2028–2063 (2028–2032)	Yes
8			No
9	MRI-ESM2-0 at GWL2	2030–2065 (2030–2034)	Yes
10			No
11	NorESM2-MM at GWL2	2025–2060 (2025–2029)	Yes
12			No
13	GFDL-ESM4 at GWL4	2059–2094 (2059–2063)	Yes
14			No
15	MRI-ESM2-0 at GWL4	2051–2086 (2051–2055)	Yes
16			No
17	NorESM2-MM at GWL4	2049–2084 (2049–2053)	Yes
18			No

Table 1.
List of models (with and without irrigation) forced with CMIP6 climatic datasets.

Crop type	Irrigation demand category	Cropping calendar		Irrigation duty	
		Season 1	Season 2	Season 1	Season 2
Lowland rice	Rice seedlings watering	1st to 20th March	1st to 20th August	120 mm per season [59]	
	Land preparation for transplanting	15th to 21st March	15th to 21st August	200 mm per season [60]	
	Net/gross irrigation demand	25th March to 3rd July	25th August to 3rd December	Net irrigation was calculated using AquaCrop software; gross irrigation was calculated using an overall irrigation efficiency of 60% for CF and 78% for AWD.	
Polycrop (beans)	Net/gross irrigation demand	1st March to 17th June	1st August to 17th November	Net irrigation was calculated using AquaCrop software; gross irrigation was calculated using an overall irrigation efficiency of 60%.	

Table 2.
Irrigation demand categories and schedules for rice and polycrops.

seasons. Given that supplementary irrigation (irrigated agriculture aligned to rainy seasons) minimises irrigation water use, thus lowering the cost of management and operation of irrigation systems, this study's cropping calendar was adapted to the rainy seasons (**Table 2**). Polycrops (represented as beans in this study) are grown from March to June or August to November and lowland rice from March to July or August to December.

Beans are dry-seeded. Thus, the only irrigation demand category is the daily gross irrigation requirement (I_{gross}) from planting to maturity. I_{gross} was calculated from the net irrigation requirement (I_{net}) using Eq. (1), where E_{ff} = 60% [45] is the scheme's overall irrigation efficiency. The daily net irrigation was estimated using the field-scale crop growth model, AquaCrop version 7 [61, 62], with soil moisture maintained at field capacity. Calibrated for various crops (including rice and beans), AquaCrop calculates net irrigation while accounting for rainfall and soil moisture balance [63]. Its modelling principle is based on the fundamental connection between water transpiration and carbon assimilation. It simulates the interactions between the plant and the soil, with the area immediately above the plant surface and below the topsoil (i.e. root zone) as upper and lower model boundaries. The plant extracts water and nutrients from the root zone depending on field management (e.g. soil fertility, soil salinity stress, field slope, contouring, terracing, method of planting, bunds, soil covering, irrigation, and practices affecting infiltration). The upper boundary links the plant-soil system to the atmosphere, determining rain input and evaporative demand (ETo) and supplying CO_2 and energy for crop growth. The Curve Number method determines the amount of rainwater lost as runoff. At the lower boundary, water drains from the system to the subsoil and the groundwater table. If the groundwater table is shallow, water can move upward into the root system by capillary rise.

$$I_{gross} = \frac{I_{net}}{E_{ff}} \tag{1}$$

The irrigation demand categories for rice include supply to nursery beds (*Nur*), field puddling requirement for transplanting rice seedlings (*SAT*), and the requirement for replenishing soil moisture from the time of transplant to maturity (I_{net}/E_{ff}), Eq. (2).

$$I_{grossRice} = Nur + SAT + \frac{I_{net}}{E_{ff}} \quad (2)$$

Rice seedlings in nurseries occupy roughly 10% of the area designated for transplantation and mature approximately 15–20 days after sowing [64]. The seasonal demand for nurseries, estimated at 120 mm [59], was applied over 20 days (1st to 20th March or 1st to 20th August) at a 6 mm/day peak rate. The daily nursery demand was estimated as the deficit between the peak rate and the average daily field water level, which was obtained using a parallel SHETRAN model with bunds but no irrigation.

The seasonal field puddling demand is roughly 150–250 mm [65], averaging 200 mm. This was applied over 7 days (15th to 21st March or 15th to 21st August) at a 28 mm/day peak rate. Like the nursery demand, the daily puddling demand was estimated as the deficit between the peak rate and the average daily field water level.

Like beans, the daily net irrigation for rice was estimated using AquaCrop, with soil moisture maintained at field capacity. Rice bunds were set in AquaCrop to account for ponded water. The gross irrigation was calculated from the net irrigation using the overall irrigation efficiency (E_{ff}) of 60% [45] for CF. Since AWD lowers infield water use by up to 30% [66], its E_{ff} was adjusted from 60 to 78%.

Table 3 shows the estimated mean annual irrigation demand in the Naigombwa catchment under the various warming levels and irrigation methods. Overall, irrigation demand increases with global warming due to increases in evapotranspiration [67].

2.3 Assessment of hydrological alteration

The indicators of hydrologic alteration (IHA), proposed by Richter et al. [26], assess the effects of flow regime change by analysing inter- and intra-annual variability in daily hydrologic data in pre- and post-impact periods. The IHA contains 33 hydrological indices subdivided into five groups that define the magnitude, frequency, duration, timing, and rate of change of flows (**Table 4**). This study calculated the hydrological indices using the IHA software (version 7.1) developed by the Nature Conservancy [68]. The software is programmed to do both parametric and nonparametric analyses. Unlike the parametric method, which assumes that the data conforms to a particular distribution, the nonparametric approach does not rely on any assumption about the underlying distribution of the data. Hydrological data are often strongly non-normal [69]. The study, therefore, used the nonparametric approach.

BL		GWL2 (relative change)		GWL4 (relative change)	
CF	AWD	CF	AWD	CF	AWD
1357 mm	1145 mm	1368 mm (0.8%)	1153 mm (0.7%)	1446 mm (6.6%)	1213 mm (6.0%)

Table 3.
Thirty-year mean annual irrigation demand at baseline (BL) and global warming levels 2 (GWL2) and 4 (GWL4). Results are based on an ensemble average.

IHA group/parameter	Regime characteristics	Number of indicators
Group 1: Magnitude of monthly water conditions	Magnitude and timing	12
Group 2: Magnitude and duration of annual extreme water conditions	Magnitude and duration	12
Group 3: Timing of annual extreme water conditions	Timing	2
Group 4: Frequency and duration of high and low pulses	Frequency and duration	4
Group 5: Rate and frequency of water condition changes	Rates of change and frequency	3

Table 4.
The 33 IHA indices used to study flow regime alterations (from [68]).

Thirty years of data were used for each pre- and post-impact period to account for natural climatic variability, given that at least 20 years of data are required [68]. The post-impact periods were targeted to the 2 and 4°C warming levels, with the baseline period (1983–2012) as the pre-impact.

2.3.1 Estimation of the degree of alteration

Like the IHA indices, the RVA metrics are programmed in the Nature Conservancy's IHA software [68]. The RVA categorises all the pre-impact data of each IHA metric into three categories [32, 33]. These include low range, middle range, and high range. In nonparametric IHA analysis, the boundaries between the RVA categories are based on quartile values [68]. These were set to the 25th and 75th quartiles to maintain IHA metrics within the middle category in 50% of the years.

The degree of alteration (D) of each of the 32 IHA metrics in the post-impact data is measured by the extent to which the RVA target range was not attained. The degree of alteration is expressed as [33]:

$$D_n = \frac{N_{on} - N_{en}}{N_{en}} \times 100\% \tag{3}$$

Where n is the nth IHA metric, and N_{on} and N_{en} represent the count of years of the IHA indices falling within the RVA target range (i.e. the middle category) in the post- and pre-impact periods, respectively. A positive value indicates an increase in the frequency of the flow metric values in the category from pre-impact to post-impact. In contrast, a negative value indicates a reduction in the frequency of values. Richter et al. [33] set thresholds for D. Values less than or equal to 33% signify slight alteration, 34 to 67% represent moderate alteration, and greater than 67% are highly altered.

The degree of alteration of a single metric (D) does not reflect the overall change in the flow regime. Thus, the overall degree of alteration was calculated for each IHA group and all IHA metrics, as expressed below [37].

$$D_o = D_{on} = \sqrt{\frac{\sum_{n=1}^{N} D_n^2}{N}} \times 100\% \tag{4}$$

Where D_o and D_{on} are the overall degree of alteration for each group and all metrics, respectively, and N is the number of IHA metrics. Following Reference [37], the overall degree of alteration is classified into five classes, including slight alteration (<20%), low alteration (20–40%), moderate alteration (40–60%), high alteration (60–80%), and severe alteration (>80%).

The outputs of the models summarised in **Table 1** were used in estimating the degree of alteration (D, Do, and D_{on}) due to climate change alone (i.e. climate-induced alterations) and the combined effects of irrigation and climate change (i.e. irrigation- and climate-induced alterations). To estimate the influence of irrigation alone (i.e. irrigation-induced alteration), climate-induced D was subtracted from the corresponding climate- and irrigation-induced D. The overall degree of alteration (D_o and D_{on}) was then calculated using Eq. (4).

3. Results

3.1 Overall hydrological prediction of the effects of irrigation and climate change on wetland inflow

The 30-year wetland inflow hydrographs at the baseline (BL) and each of the two global warming levels (GWL2 and GWL4) are shown in **Figures 3** and **4**, respectively. Each figure includes hydrographs for conditions with CF and AWD and without irrigation at BL and GWL2 or GWL4. It can be seen that extreme flows (flood and low flows) are the most affected by irrigation. However, the difference between the irrigation-induced hydrographs (CF and AWD) is negligible for a particular warming level. The median daily flow at BL under CF and AWD is 1.81 and 1.82 m^3/s, respectively. Similarly, the median flows at GWL2 are 2.15 and 2.09 m^3/s, while those at GWL4 are 2 and 2.02 m^3/s.

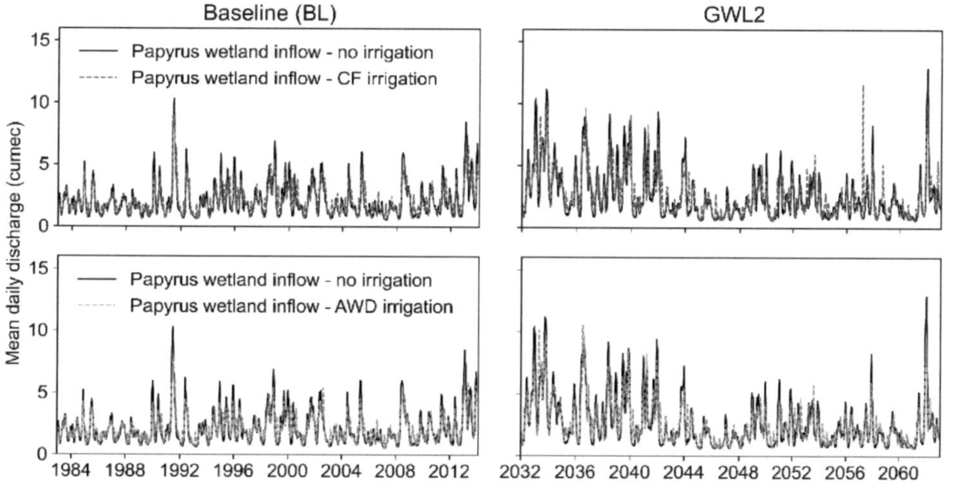

Figure 3.
Papyrus wetland inflow for scenarios with and without irrigation at baseline (BL) and global warming level 2 (GWL2). The top row is model runs with continuous flooding (CF) irrigation. The bottom row is model runs with alternate wetting and drying irrigation.

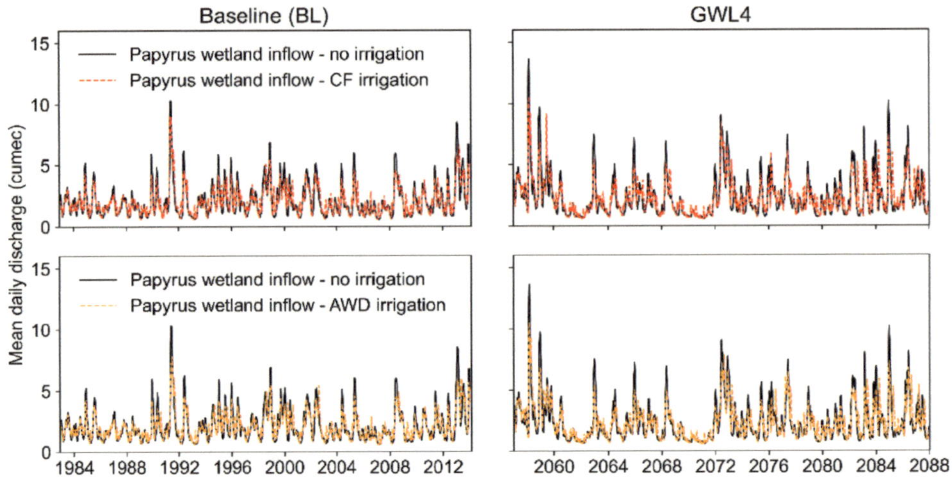

Figure 4.
Papyrus wetland inflow for scenarios with and without irrigation at baseline (BL) and global warming level 4 (GWL4). The top row is model runs with continuous flooding (CF) irrigation. The bottom row is model runs with alternate wetting and drying (AWD) irrigation.

The median daily flow for the no irrigation condition is 1.66, 1.9, and 1.73 m^3/s at BL, GWL2, and GWL4, respectively. These values are slightly lower than the corresponding values for the conditions with irrigation by at least 8, 9 and 13% at BL, GWL2, and GWL4, respectively. The increased median daily flows for the conditions with irrigation could be explained by the fact that lowland rice irrigation enhances river flow [70] due to increased groundwater recharge [70, 71].

Table 5 shows the irrigated sites' median annual water balance (precipitation, evapotranspiration, and percolation/runoff) for the various warming levels and irrigation scenarios. Similar to findings in wetland-dominated catchments [72] and other agroecosystems [67], evapotranspiration increases with warming. However, for a particular warming level (BL, GWL2, or GWL4), the evapotranspiration rates are similar irrespective of the irrigation scenario.

Water balance component	Baseline (BL)		GWL2 (relative change, %)		GWL4 (relative change, %)	
	CF irrigation	AWD irrigation	CF irrigation	AWD irrigation	CF irrigation	AWD irrigation
Precipitation (mm)	1313	1313	1349 (2.7)	1349 (2.7)	1402 (6.5)	1402 (6.5)
Irrigation water (mm)	1357	1145	1368 (0.7)	1153 (0.7)	1446 (6.9)	1213 (6.3)
Total evapotranspiration (mm)	907	907	912 (0.3)	912 (0.3)	965 (6.1)	965 (6.1)
Percolation and runoff (mm)	1762	1550	1805 (2.4)	1590 (2.6)	1883 (7.1)	1651 (6.6)

Table 5.
Median annual water balance at the irrigated sites at BL, GWL2, and GWL4. Results are based on an ensemble average.

3.2 Effects of future climate on the Naigombwa wetland inflow regime

The projected degree of alteration of each IHA group (D_o) and all IHA metrics (D_{on}) for the conditions with and without irrigation at the baseline are summarised in **Table 6**. It can be seen that climate-induced D_o and D_{on} (columns 3 and 4) are generally dependent on the baseline condition. As expected, they are more prominent for the irrigated baseline, except for Group 3 metrics at GWL4. Climate-induced D_{on} values of 67.1 and 63.4%, indicating high alteration, are projected at GWL2 and GWL4, which are more significant than the moderate alterations of 56.2 and 53.6% for the no-irrigation baseline. Similarly, climate-induced D_o at GWL2 ranges from moderate (56.7%) to severe (100%) for the irrigated baseline and slight (19.8%) to severe (97.1%) for the unirrigated baseline. However, at GWL4, D_o ranges from low to severe irrespective of the baseline condition (i.e. 30.9–100% and 32.2–96.2%).

Overall, for the irrigated baseline, all groups are significantly affected (moderate-to-severe alteration) irrespective of the warming level, except for Group 3 (timing of annual extreme water conditions) at GWL4. Similarly, most groups are significantly altered for the no-irrigation baseline, except for Group 3 at GWL2 and GWL4, Group 1 (magnitude of monthly water conditions) at GWL4, and Group 2 (magnitude and duration of annual extreme water conditions) at GWL4.

3.3 Effects of irrigation on the Naigombwa wetland inflow regime

The projected irrigation-induced degrees of alteration (D_o and D_{on}) are summarised in **Table 6** (columns 7 and 8). D_{on} values are generally insignificant, ranging

IHA group	Baseline condition	Climate-induced		Climate- and irrigation-induced		Irrigation-induced	
		GWL2	GWL4	GWL2	GWL4	GWL2	GWL4
Degree of alteration of each IHA group, D_o (%)							
Group 1	No irrigation	41.8	37.3	52.1	43.2	29.1	23.3
	With irrigation	57.2	47.8	46.7	42.5	21.3	22.4
Group 2	No irrigation	48.1	39.4	44.6	37.1	31.6	21.2
	With irrigation	56.7	55.3	59.9	42.5	29.1	21.7
Group 3	No irrigation	19.8	32.2	29.6	39.5	23.8	9.9
	With irrigation	58.1	30.9	32.2	39.5	31.9	18.2
Group 4	No irrigation	97.1	96.2	98.1	93.9	2	7
	With irrigation	100	100	99	97.8	2.1	4.5
Group 5	No irrigation	75.4	82.2	59.7	72.9	27.7	18.4
	With irrigation	89.3	94.2	83.1	88.1	11.4	8.1
Degree of alteration of all IHA metrics, D_{on} (%)							
All IHA metrics	No irrigation	56.2	53.6	57	53.2	27.9	20.1
	With irrigation	67.1	63.4	63	57.2	23.4	19.5

Table 6.
Thirty-year overall degree of alteration for baseline conditions with and without irrigation. Results are based on an ensemble average.

from slight to low alteration (19.5–27.9%), irrespective of the baseline condition and warming level. Similarly, D_o values are negligible (slight to low alterations of 2–31.9%). The D_o and D_{on} values are much lower than the corresponding climate-induced degrees of alteration described in the previous section. This indicates that climate change will be more prominent in altering wetland inflows in the Nigombwa catchment for the irrigation area considered.

3.4 The combined effects of irrigation and future change on wetland inflow

Table 6 (columns 5 and 6) summarises the climate- and irrigation-induced overall degrees of alteration (D_o and D_{on}). The alterations' significances (low to severe at GWL2 (29.6–99%) and GWL4 (37.1–97.8%)) are generally similar to those of the climate-induced alterations (slight to severe at GWL2 (19.8–100%) and low to severe at GWL4 (30.9–100%)). The "effects of future climate on the Naigombwa wetland inflow regime" section describes the climate-induced alterations. As indicated by D_o, the most altered metrics (moderate-to-severe alteration) by the combined effects of irrigation and climate change at GWL2 for the no irrigation baseline are in Group 4 (severe alteration, 98.1%) and Groups 1, 2, and 5 (moderate alterations of 52.1, 44.6, and 59.7%). Similarly, Group 4 (severe alteration, 97.8%), Group 5 (high alteration, 72.9%), and Group 1 (moderate alteration, 43.2%) are the most altered at GWL4. Meanwhile, groups 4 and 5 (severe alteration of 83.1 to 99%) and groups 1 and 2 (moderate alteration of 42.5–59.9%) are the most altered for the irrigated baseline, irrespective of the warming level.

Thus, given that the climate- and irrigation-induced alterations (D_o and D_{on}) capture the influences of irrigation and climate change and are more pronounced than the irrigation-induced alterations, the analysis of each IHA metric's projected alteration (D) in the following results subsections focuses on the climate- and irrigation-induced modelling output.

3.5 Effects of irrigation and climate change on the frequency and duration of high and low pulses

The projected climate- and irrigation-induced degree of alteration of each IHA metric (D) for the condition with irrigation at the baseline is summarised in columns 9 and 10 of **Table 7** in the Appendix. The corresponding D values for the condition of no-irrigation at the baseline are in **Table 8** of the Appendix. D values (−87 to −100%, denoting highly altered) in the most altered group (Group 4) indicate a future highly modified frequency and duration of high and low pulses, irrespective of the baseline condition and warming level. The effect is fewer high and low pulses but a much longer duration when they occur. The changes in the frequency and duration of the high and low pulses are shown in columns 5 and 6, with the baseline values in column 2.

The negative D values also indicate an increased occurrence of low and high pulses outside the domain of the baseline middle-range (i.e. the 25th and 75th quartiles of the baseline period). This can be seen in **Figure 5**, which shows the annual count of pulses and the median annual event duration for the baseline period (1983 to 2012), GWL2 (2032 to 2061), and GWL4 (2057 to 2086). Pulse counts will increase in the low range (0–25th quartile), with the duration of occurrence falling more often in the high range (75th–100th quartile).

Overall, the 30-year median annual low pulse frequency is expected to drop from about 22 and 17 for baseline conditions with and without irrigation, respectively, to

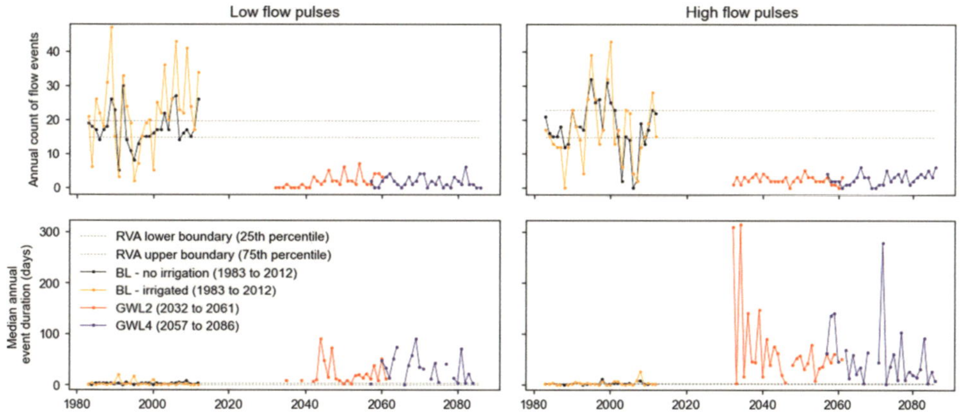

Figure 5.
Alterations in the annual frequency (top row) and median yearly duration (bottom row) of low (left column) and high (right column) flow pulses due to the combined effect of irrigation and climate change.

2 at GWL2 and 2 and 1 at GWL4. Similarly, the 30-year median annual high pulse frequency is expected to decline from about 15 and 18 to 2 at GWL2 and 3 at GWL4. Meanwhile, the 30-year median annual low pulse duration will increase from 2 and 3 days, respectively, to 24 and 12 days at GWL2 and 25 and 32 days at GWL4. The high pulse duration will increase from 2 to 45 and 35 days at GWL2 and GWL4, respectively. Although evapotranspiration increases with warming [67, 72], increased precipitation and reduced runoff can offset its effect by alleviating soil water stress [73]. Projections show increasing rainfall with warming over the East African region [74–76]. This could be the reason for the negligible difference in alterations at the 2 and 4°C warming levels.

3.6 Effects of irrigation and climate change on the rate and frequency of water condition changes

Group 5, the second most altered IHA group by the combined effects of irrigation and climate change, quantifies changes in the rise/fall rates of high flows, including the number of reversals in the flow conditions. Flow reversals are related to the frequency of events [77], which the Group 4 metrics have captured, as explained in the previous section.

As seen in the previous section, the combined effects of irrigation and climate change increase the duration of high pulses. This is likely due to the projected decline in high flow rise and fall rates (**Figure 6**). Although lowland rice irrigation enhances river flow through seepage [70], which can reduce high flow rise/fall rates due to slow flux rates of groundwater systems, the irrigation command area in this study has a negligible effect compared to future climate change, as explained in the "effects of irrigation on the Naigombwa wetland inflow regime" section. Global climate models (GCMs) commonly suffer from too much light rain with few heavy showers, yet in reality, light-rain events are low in frequency in the tropics, with moderate-to-heavy rains often occurring [78]. This can result in a decline in high flow rise/fall rates in catchment models. The effect significantly alters the high flow rise and fall rates at the inlet to Naigombwa wetlands, with D ranging from −43.8 to −93.8% (denoting moderate to highly altered) for the irrigated baseline (**Table 7** in the Appendix). Similarly, for the no irrigation baseline (**Table 8** in the Appendix), the rise and fall rates are

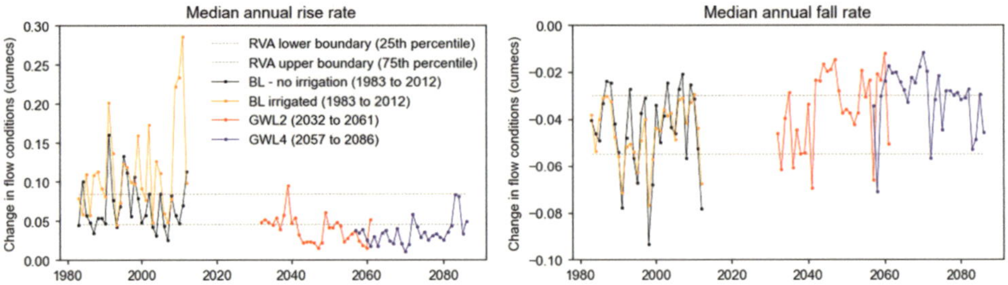

Figure 6.
Alterations in the median annual rate of change (hydrograph rise [left graph] and fall [right graph] rates) of flow conditions due to the combined effect of irrigation and climate change.

significantly altered by −37.5 to −75%, except for the fall rate at GWL2 (slight alteration of −18.8%). The negative *D* values indicate the increased occurrence of high flow rise and fall rates outside the domain of the baseline middle-range (i.e. the 25th and 75th quartiles of the baseline period), with increased occurrence in the low-range (0–25th quartile, **Figure 6**).

Overall, the 30-year median rise rate is predicted to decline from 0.1 and 0.06 m^3/s at BL for the condition with and without irrigation to 0.04 and 0.03 m^3/s at GWL2 and GWL4, respectively. Similarly, the 30-year median fall rate will drop from 0.04 m^3/s (for the conditions with and without irrigation) to 0.04 and 0.03 m^3/s.

3.7 Effects of irrigation and climate change on the magnitude of monthly water conditions

As the third most altered IHA group, the magnitude of monthly flows in Group 1 is significantly transformed (moderately or highly modified) in about half of the months, irrespective of the baseline (BL) condition and warming level. *D* ranges from −43.8 to −93.8% and − 37.5 to −87.5% at GWL2 for the no irrigation and irrigated baselines (**Tables 8** and 7, respectively, in the Appendix). Similarly, *D* at GWL4 ranges from −37.5 to −93.8% and − 37.5 to −68.8%. The affected months at GWL2 are March, April, August, September, November, December, and January for the no-irrigation BL and March, April, July, December, and January for the irrigated BL (**Figure 7**). The corresponding periods at GWL4 are March, April, July, November, December, and February, and March, April, and November to February. The least affected months for all conditions are May and June. The negative *D* values indicate a projected decline in the frequency of median monthly wetland inflows in the middle range of the baseline period (25th–75th quartiles), with a significant increase in the high range (75th–100th quartile, **Figure 7**).

Although GCMs generally predict increased future rainfall over the East African region [74–76], thus the projected increase in median monthly flows in the high range, changes are more significant during the second rainy season (August–November) compared to the first (March–May) [79]. This explains the insignificant flow alterations in the first dry season (May–June) compared to the second (December to February). Overall, the 30-year median monthly flows will increase from 0.86–2.71 to 0.97–2.08 m^3/s at BL with and without irrigation, respectively, to 1.27–3.33 m^3/s at GWL2 and 1.39–2.53 m^3/s at GWL4 (**Tables** 7 and **8** in the Appendix).

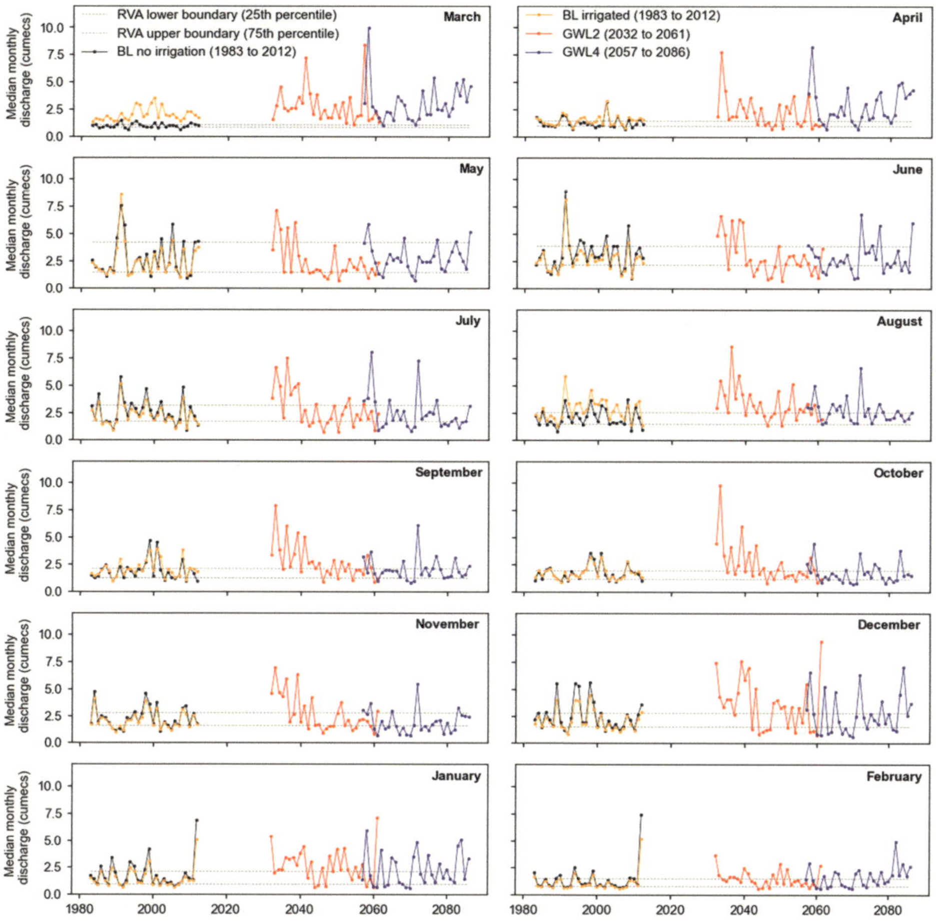

Figure 7.
Alterations in the median monthly flow magnitude due to the combined effect of irrigation and climate change. The water year in the study area is from March (top left graph) to February (bottom right graph).

3.8 Effects of irrigation and climate change on annual extreme water conditions

The predictions of the degree of alteration (D) of the magnitude of annual extreme flows in the Group 2 metrics indicate dependence on the baseline (BL) condition. As the fourth most altered IHA group, the Group 2 metrics are more often significantly transformed (moderately or highly) for the BL condition with irrigation. All annual extremes (1-, 3-, 7-, 30-, and 90-day minimum/maximum flows) are changed considerably at all warming levels for the irrigated BL, except for the 30-day minimum flow at GWL2 (**Table 7** in the Appendix). Meanwhile, significant modifications for the no irrigation BL occur only in the 1-, 3-, 7-, and 90-day minimum flows and the 1-day maximum flow at GWL2 and GWL4, including the 90-day maximum flow at GWL2 (**Table 8** in the Appendix). The D values are projected to range from −50 to −93.8% and − 37.5 to −56.3% at GWL2 for the conditions with and without irrigation at BL, respectively. Similarly, projections at GWL4 indicate D in the range of −37.5 to −62.5% and − 50 to −56.3%. The negative

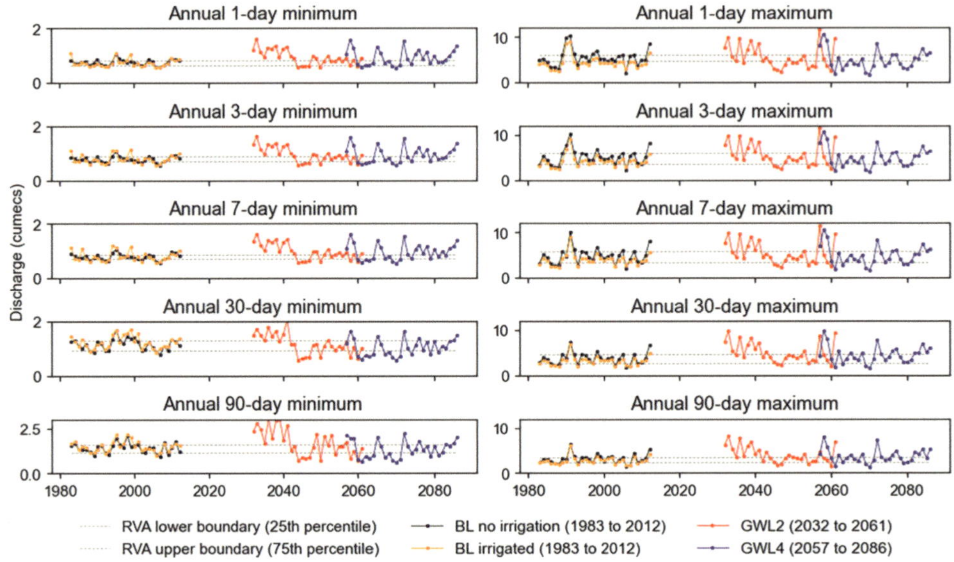

Figure 8.
Alterations in the median annual extreme flows (1-, 3-, 7-, 30-, and 90-day minimum (left column) and maximum (right column) values) due to the combined effect of irrigation and climate change.

D values indicate a projected decline in the frequency of median annual extremes in the baseline period's middle range (25th–75th quartiles), with an increase in the high range (75th–100th quartile, **Figure 8**).

Overall, the 30-year median annual minimum discharge is predicted to increase from 0.71–1.46 to 0.74–1.42 m^3/s at BL with and without irrigation, respectively, to 0.89–1.51 m^3/s at GWL2 and 0.84–1.29 m^3/s at GWL4 (**Tables** 7 and **8** in the Appendix). Similarly, the median annual maximum flow will increase from 2.63–4.24 and 2.9–5.2 m^3/s to 3.5–4.8 and 3.3–4.6 m^3/s.

4. Discussion

Findings in this study show that climate change will play a much more significant role than irrigation in influencing future wetland inflow in the Naigombwa catchment for the target lowland rice acreage of 2200 ha. The highland and lake regions in East Africa, such as the Mpologoma River basin, are expected to experience precipitation increases [74–76], with mean annual precipitation increases of 9.2 and 25.7% in the near (2030–2059) and far (2070–2099) future, respectively [76]. The projected increase in precipitation is significantly higher than the predicted increase in irrigation requirement in the Naigombwa catchment. Mean annual irrigation demand under continuous flooding (CF) is estimated to increase only by 0.8 and 6.6% at the 2 and 4°C warming levels, respectively (**Table 3**).

The unmistakable evidence of climate change, including elevated temperatures and shifts in precipitation patterns, is already evident, leading to tangible impacts on ecosystems, biodiversity, and human populations [80]. In the case of wetland-dominated catchments, climate change affects ecosystem services that communities primarily depend on. The risks to ecosystem services depend on the magnitude and rate of warming, geographic location, levels of development and vulnerability, and the choices and implementation of adaptation and mitigation options [81]. Increased temperature impacts ecosystems by shifting species' distributional range and phenological events (i.e. the timing of recurring seasonal events) [80]. Plants and animals tend to move to higher altitudes as temperatures get warmer [82, 83]. However, species that may be unable to keep pace with the changing climate lag behind, leading to their eventual extinction [80]. The threat of extinction is more potent for protected areas such as wetlands, as they hold many species with small distributional ranges [84].

A number of IHA-RVA studies on the effects of irrigation water abstraction on river systems show that extreme flows are most altered [35–39]. Although based on historical periods, the findings of these studies are generally similar to those in this study. In the Naigombwa catchment, extreme flows are significantly altered in the frequency and duration of low and high flows, the rate of change of flow conditions (i.e. hydrograph rise and fall rates), and the magnitude of annual extremes. These corroborate the findings by Pompeu et al. [85] and Stefanidis et al. [36] in Europe, Xue et al. [37] and Wang et al. [39] in China, and Yasarer et al. [38] in the United States, who evidenced significant changes in the magnitude, frequency, duration, and rate of change of extremes.

Findings in this study also show that irrigation applied to supplement rainwater (i.e. when irrigated crop lifecycle is aligned to rain seasons) increases monthly river flow during rain periods but lowers flows during low rain/dry periods (**Figure 9**). The crop cycle in the Naigombwa catchment is from March to June/July for the first season and August to November/December for the second season. **Figure 9** shows that irrigation generally enhances wetland inflow from March to May and August to October, while flows are lowered from June to July and December to February, corresponding to the first and second low rain/dry seasons in a water year. Pompeu et al. [85] witnessed similar changes in a catchment irrigated with dam water in Spain. Summer flows significantly increased while winter flows declined. Pompeu et al. [85] concluded that irrigation alters and can even shift seasonal flow patterns. However, this will depend on the source water (surface water or groundwater), irrigation period, the irrigation method, and the irrigation command area.

Africa's population, at over 1.4 billion, is projected to reach nearly 2.5 billion by 2050, equivalent to 25% of the world's population. The share will increase to approximately 40% by the end of the century [86]. Increasing populations imply increased demand for food, which, in the case of Africa, can be met through intensive agriculture, including irrigation [12, 15]. This study, limited to the irrigation acreage of 2200 ha, shows that the effects of climate change on wetlands will be much more

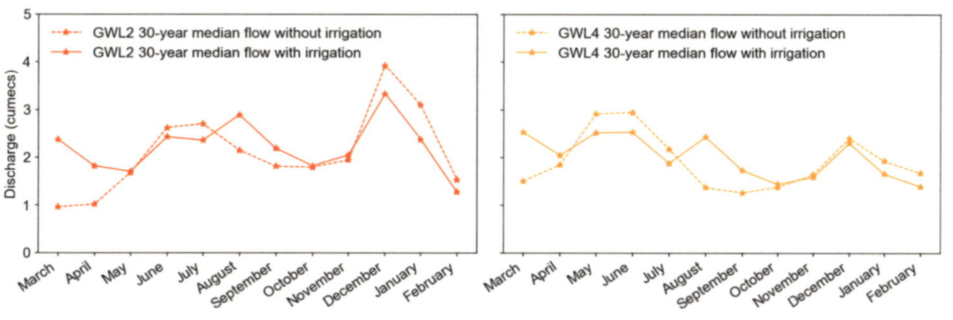

Figure 9.
Thirty-year median monthly wetland inflow for situations with and without irrigation at GWL2 (left column) and GWL4 (right column).

dominant than irrigation. However, due to the projected increasing population on the African continent, societal pressure on wetlands (e.g. river regulation and diversion for irrigation [8, 87]) will increase. For example, the heads of states' African Development Agenda 2063 call for increased irrigation water by 60% from the 2013 levels. Since catchments have unique characteristics (such as land cover, lithology, precipitation, and evapotranspiration), the effects of societal pressure on wetlands are best assessed on a case-by-case basis. In the case of the Naigombwa catchment, projections of the irrigation command area are recommended to better predict the effects of lowland rice irrigation on the papyri wetland flow regime under a future climate.

It is worth noting that the findings in this study are subject to limitations and uncertainties of the approaches taken. Firstly, AquaCrop and SHETRAN use soil-saturated hydraulic conductivity (Ksat) and soil water content at saturation (SAT) as parameters. As a field-scale model, Ksat and SAT values for AquaCrop are often determined experimentally [88]. Due to errors in input data (e.g. precipitation, measured water level/river flow, soil, and land use/cover), in addition to assumptions in model physics and structure, the calibrated parameters of catchment-scale models such as SHETRAN usually differ from experimental values [89–91]. Thus, this study used the Ksat and SAT field-scale values of [92] in AquaCrop. The values were, however, adjusted in SHETRAN to attain a working model.

Secondly, this study forced the calibrated SHETRAN model for the Mpologoma River basin (see ref. [53]) with future climate data (CMIP6 rainfall and potential evapotranspiration products). The model was calibrated with the MSWEP rainfall data, a satellite-based precipitation product (SPP). SPPs have inherent errors that can lead to uncertainty in hydrological model parameter estimation and flow simulation [93, 94]. Thirdly, future climate data generally contribute the most significant uncertainty to hydrologic studies [95], with the variability of GCM projections as the most critical contributor to flow prediction uncertainty in catchments with wetlands [96]. This study used only three GCMs out of a possible nine due to limitations in resolution. Thus, using a larger GCM or regional climate model ensemble to lower prediction uncertainty is recommended.

5. Conclusions

This study identified and quantified the alterations in the flow regime of a papyrus-dominated wetland system due to the effects of global warming and irrigated agriculture. By calculating indices that define the degree of alteration between the hypothetical and the current (baseline) scenario, the impacts of lowland rice irrigation amid climate change in the Naigombwa catchment in Uganda are quantified. Results indicate that climate change, due to the projected increase in precipitation over the study area, will significantly influence wetland inflow more than irrigation for the irrigation command area of 2300 ha, equivalent to the irrigated area to catchment area ratio of order 1:70. Similar to studies in other parts of the world, findings show that extreme flows are most affected by water diversion for agriculture.

High (flood) flows are critical for floodplain-dominated papyrus wetlands. In the Naigombwa catchment, high flows are negatively affected (i.e. reduced) in frequency, peak amplitude, and rate of change. This, ecologically, will affect papyrus sexual reproduction by limiting opportunities for soil rewetting, which is critical for seed germination and establishment. Papyrus is resilient to limitations in sexual reproduction by regenerating asexually; however, this will curtail the exchange of genetic material and promote genetic diversity, which is only possible through sexual reproduction. The consequence is a weaker potential to adapt, through genetic interaction, to environmental pressures associated with climate warming. Studies show that wetlands, one of the richest ecosystems in the world, will be adversely affected by global warming through temperature increases [4] and increased water abstraction for food production [8]. Thus, the findings in this study can help catchment managers in similar ecosystems plan for conservation programmes that can enhance the productivity of papyrus plants, given that it is a keystone species on which wetland biodiversity depends.

Acknowledgements

The UK Commonwealth Scholarship funded this work for the first author.

Conflict of interest

The authors declare no conflict of interest.

Appendix

IHA metric	The median value of the IHA metric					Degree of alteration, D (%)					
	Baseline, BL (with irrigation)	Climate-induced		Climate- and irrigation-induced		Climate-induced		Climate- and irrigation-induced		Irrigation-induced	
		GWL2	GWL4	GWL2	GWL4	GWL2	GWL4	GWL2	GWL4	GWL2	GWL4
Group 1: Magnitude of monthly water conditions											
March	1.85	0.961	1.509	2.374	2.525	−87.5	−68.8	−50	−62.5	37.5	6.3
April	1.391	1.016	1.846	1.819	2.048	−81.3	−56.3	−68.8	−68.8	12.5	−12.5
May	1.908	1.679	2.907	1.701	2.517	−12.5	0	12.5	25	25	25
June	2.638	2.617	2.944	2.425	2.526	−43.8	−31.3	−18.8	−18.8	25.0	12.5
July	2.036	2.696	2.177	2.352	1.878	−37.5	−50	−37.5	−31.3	0	18.8
August	2.711	2.15	1.37	2.884	2.427	−18.8	−50	−12.5	0	6.3	50
September	1.859	1.81	1.267	2.188	1.729	−6.3	−62.5	−31.3	−18.8	−25.0	43.8
October	1.548	1.793	1.378	1.821	1.44	−31.3	−25	−25	−18.8	6.3	6.3
November	1.798	1.942	1.641	2.048	1.587	−31.3	−37.5	−31.3	−37.5	0	0
December	1.773	3.923	2.404	3.328	2.3	−75	−56.3	−87.5	−62.5	−12.5	−6.3
January	1.23	3.089	1.926	2.377	1.658	−93.8	−50	−75	−43.8	18.8	6.3
February	0.863	1.535	1.673	1.272	1.394	−68.8	−43.8	−31.3	−56.3	37.5	−12.5
Group 1 overall degree of alteration, D_o (%)						57.2	47.8	46.7	42.5	21.3	22.4
Group 2: Magnitude and duration of annual extreme water conditions											
1-day minimum	0.706	0.812	0.82	0.889	0.835	−43.8	−56.3	−93.8	−43.8	−50	12.5
3-day minimum	0.746	0.816	0.824	0.9	0.844	−25	−25	−68.8	−43.8	−43.8	−18.8
7-day minimum	0.768	0.823	0.835	0.942	0.882	−25	−18.8	−68.8	−50	−43.8	−31.3
30-day minimum	1.205	0.851	0.899	1.133	1.003	−75	−56.3	−31.3	−37.5	43.8	18.8
90-day minimum	1.463	1.188	1.096	1.51	1.291	−75	−50	−50	−43.8	25	6.3
1-day maximum	4.236	5.303	5.12	4.759	4.573	−75	−68.8	−68.8	−62.5	6.3	6.3

IHA metric	The median value of the IHA metric					Degree of alteration, D (%)					
	Baseline, BL (with irrigation)	Climate-induced		Climate- and irrigation-induced		Climate-induced		Climate- and irrigation-induced		Irrigation-induced	
		GWL2	GWL4	GWL2	GWL4	GWL2	GWL4	GWL2	GWL4	GWL2	GWL4
3-day maximum	4.066	5.3	5.115	4.75	4.564	−62.5	−75	−56.3	−37.5	6.3	37.5
7-day maximum	3.907	5.28	5.098	4.709	4.544	−62.5	−68.8	−56.3	−37.5	6.3	31.3
30-day maximum	3.203	5.037	4.713	4.275	4.024	−75	−75	−62.5	−50	12.5	25
90-day maximum	2.628	4.132	3.914	3.521	3.282	−68.8	−68.8	−62.5	−50	6.3	18.8
Base flow index	0.405	0.371	0.44	0.358	0.396	−18.8	−37.5	−50	−18.8	−31.3	18.8
Group 2 overall degree of alteration, D_o (%)						56.7	55.3	59.9	42.5	29.1	21.7
Group 3: Timing of annual extreme water conditions											
Date of minimum	54	76	52	58	44	81	0	43.8	−25	−37.5	−25
Date of maximum	37	326	157	345	87	−13	44	12.5	50	25	6.3
Group 3 overall degree of alteration, D_o (%)						58.1	30.9	32.2	39.5	31.9	18.2
Group 4: Frequency and duration of high and low pulses											
Low pulse count	22	1	1	2	2	−100	−100	−100	−100	0	0
Low pulse duration	2	69	60	24	25	−100	−100	−95.8	−100	4.2	0
High pulse count	15	1	2	2	3	−100	−100	−100	−100	0	0
High pulse duration	2	84	72	45	35	−100	−100	−100	−90.9	0	9.1
Group 4 overall degree of alteration, D_o (%)						100.0	100.0	99.0	97.8	2.1	4.5
Group 5: Rate and frequency of water condition changes											
Rise rate	0.098	0.027	0.029	0.04	0.033	−100	−100	−93.8	−87.5	6.3	12.5
Fall rate	−0.044	−0.023	−0.021	−0.035	−0.028	−62.5	−81.3	−43.8	−75	18.8	6.3

IHA metric	The median value of the IHA metric					Degree of alteration, D (%)					
	Baseline, BL (with irrigation)	Climate-induced		Climate- and irrigation-induced		Climate-induced		Climate- and irrigation-induced		Irrigation-induced	
		GWL2	GWL4	GWL2	GWL4	GWL2	GWL4	GWL2	GWL4	GWL2	GWL4
Number of reversals	115	36	38	46	50	−100	−100	−100	−100	0	0
Group 5 overall degree of alteration, D_o (%)						89.3	94.2	83.1	88.1	11.4	8.1
The overall degree of alteration of all metrics, D_{on} (%)						67.1	63.4	63	57.2	23.4	19.5

Table 7.
IHA-RVA analysis over 30 years showing the degrees of climate- and irrigation-induced alterations with irrigation in the baseline. Results are based on an ensemble average.

IHA metric	The median annual value of the IHA metric					Degree of alteration, D (%)					
	Baseline, BL (no irrigation)	Climate-induced		Climate- and irrigation-induced		Climate-induced		Climate- and irrigation-induced		Irrigation-induced	
		GWL2	GWL4	GWL2	GWL4	GWL2	GWL4	GWL2	GWL4	GWL2	GWL4
Group 1: Magnitude of monthly water conditions											
March	0.973	0.961	1.509	2.374	2.525	−37.5	−50	−93.8	−93.8	−56.3	−43.8
April	1.151	1.016	1.846	1.819	2.048	−50	−43.8	−50	−75	0	−31.3
May	1.94	1.679	2.907	1.701	2.517	−12.5	12.5	25	31.3	37.5	18.8
June	2.891	2.617	2.944	2.425	2.526	−25	−6.3	−18.8	−6.3	6.3	0
July	2.199	2.696	2.177	2.352	1.878	−37.5	−37.5	−31.3	−37.5	6.3	0
August	1.703	2.15	1.37	2.884	2.427	0	−25	−50	0	−50	25
September	1.665	1.81	1.267	2.188	1.729	−6.3	−37.5	−43.8	6.3	−37.5	43.8
October	1.501	1.793	1.378	1.821	1.44	−12.5	−18.8	−25	−18.8	−12.5	0
November	1.935	1.942	1.641	2.048	1.587	−37.5	−31.3	−43.8	−37.5	−6.3	−6.3
December	2.079	3.923	2.404	3.328	2.3	−81.3	−56.3	−87.5	−37.5	−6.3	18.8
January	1.372	3.089	1.926	2.377	1.658	−68.8	−31.3	−68.8	−31.3	0	0
February	0.961	1.535	1.673	1.272	1.394	−43.8	−56.3	−6.3	−37.5	37.5	18.8
Group 1 overall degree of alteration, D_o (%)						41.8	37.3	52.1	43.2	29.1	23.3
Group 2: Magnitude and duration of annual extreme water conditions											
1-day minimum	0.743	0.812	0.82	0.889	0.835	−37.5	−31.3	−75	−50	−37.5	−18.8
3-day minimum	0.759	0.816	0.824	0.9	0.844	−37.5	−25	−68.8	−50	−31.3	−25
7-day minimum	0.773	0.823	0.835	0.942	0.882	−37.5	−31.3	−68.8	−56.3	−31.3	−25
30-day minimum	1.16	0.851	0.899	1.133	1.003	−62.5	−50	−6.3	−31.3	56.3	18.8
90-day minimum	1.416	1.188	1.096	1.51	1.291	−68.8	−50	−50	−56.3	18.8	−6.3

IHA metric	**The median annual value of the IHA metric**					**Degree of alteration, D (%)**					
	Baseline, BL (no irrigation)	**Climate-induced**		**Climate- and irrigation-induced**		**Climate-induced**		**Climate- and irrigation-induced**		**Irrigation-induced**	
		GWL2	**GWL4**	**GWL2**	**GWL4**	**GWL2**	**GWL4**	**GWL2**	**GWL4**	**GWL2**	**GWL4**
1-day maximum	5.213	5.303	5.12	4.759	4.573	−68.8	−56.3	−56.3	−50	12.5	6.3
3-day maximum	4.964	5.3	5.115	4.75	4.564	−37.5	−25	−6.3	−6.3	31.3	18.8
7-day maximum	4.753	5.28	5.098	4.709	4.544	−37.5	−25	−6.3	−6.3	31.3	18.8
30-day maximum	3.591	5.037	4.713	4.275	4.024	−50	−50	−12.5	−25	37.5	25
90-day maximum	2.899	4.132	3.914	3.521	3.282	−68.8	−56.3	−37.5	−31.3	31.3	25
Base flow index	0.413	0.371	0.44	0.358	0.396	−12.5	−31.3	−37.5	6.3	−25	37.5
Group 2 overall degree of alteration, D_o (%)						48.1	39.4	44.6	37.1	31.6	21.2
Group 3: Timing of annual extreme water conditions											
Date of minimum	66	76	52	58	44	25	−12.5	37.5	−25	12.5	−12.5
Date of maximum	37	326	157	345	87	−12.5	43.8	18.8	50	31.3	6.3
Group 3 overall degree of alteration, D_o (%)						19.8	32.2	29.6	39.5	23.8	9.9
Group 4: Frequency and duration of high and low pulses											
Low pulse count	17	1	2	2	1	−100	−100	−100	−100	0	0
Low pulse duration	3	76	40	12	32	−88	−84	−92	−88	−4	−4
High pulse count	18	1	2	2	3	−100	−100	−100	−100	0	0
High pulse duration	2	84	72	45	35	−100	−100	−100	−87	0	13.3
Group 4 overall degree of alteration, D_o (%)						97.1	96.2	98.1	93.9	2	7
Group 5: Rate and frequency of water condition changes											
Rise rate	0.057	0.027	0.029	0.04	0.033	−56.3	−81.3	−37.5	−75	18.8	6.3
Fall rate	−0.043	−0.023	−0.021	−0.035	−0.028	−62.5	−68.8	−18.8	−37.5	43.8	31.3
Number of reversals	88	36	38	46	50	−100	−94.4	−94.4	−94.4	5.6	0
Group 5 overall degree of alteration, D_o (%)						75.4	82.2	59.7	72.9	27.7	18.4

DOI: http://dx.doi.org/10.5772/intechopen.1007716

IHA metric	The median annual value of the IHA metric					Degree of alteration, D (%)					
	Baseline, BL (no irrigation)	Climate-induced		Climate- and irrigation-induced		Climate-induced		Climate-and irrigation-induced		Irrigation-induced	
		GWL2	GWL4	GWL2	GWL4	GWL2	GWL4	GWL2	GWL4	GWL2	GWL4
The overall degree of alteration of all metrics, D_{on} (%)						56.2	53.6	57	53.2	27.9	20.1

Table 8.
IHA-RVA analysis over 30 years showing the degrees of climate- and irrigation-induced alterations with no irrigation in the baseline. Results are based on an ensemble average.

Author details

Alem Oyarmoi*, Stephen Birkinshaw, Caspar J.M. Hewett and Hayley J. Fowler
School of Engineering, Newcastle University, Newcastle Upon Tyne, NE, UK

*Address all correspondence to: a.oyarmoi2@ncl.ac.uk

IntechOpen

© 2025 The Author(s). Licensee IntechOpen. This chapter is distributed under the terms of the Creative Commons Attribution License (http://creativecommons.org/licenses/by/4.0), which permits unrestricted use, distribution, and reproduction in any medium, provided the original work is properly cited.

References

[1] WEF. The Global Risks Report 2023. Geneva, Switzerland: World Economic Forum (WEF); 2023

[2] Pörtner HO, Roberts DC, Adams H, Adelekan I, Adler C, Adrian R, et al. Technical summary. In: Pörtner HO, Roberts DC, Tignor M, Poloczanska ES, Mintenbeck K, Alegria A, et al., editors. Climate Change 2022: Impacts, Adaptation and Vulnerability Contribution of Working Group II to the Sixth Assessment Report of the Intergovernmental Panel on Climate Change. Cambridge, UK and New York, NY, USA: Cambridge University Press; 2022. pp. 37-118

[3] Millennium Ecosystem Assessment. Ecosystems and Human Well-Being: Synthesis. Washington, DC: Island Press; 2005

[4] Salimi S, Almuktar SAAAN, Scholz M. Impact of climate change on wetland ecosystems: A critical review of experimental wetlands. Journal of Environmental Management [Internet]. 2021;**286**:112160. Available from: https://www.sciencedirect.com/science/article/pii/S030147972100222X

[5] Erwin KL. Wetlands and global climate change: The role of wetland restoration in a changing world. Wetlands Ecology and Management [Internet]. 2009;**17**(1):71-84. DOI: 10.1007/s11273-008-9119-1

[6] Reddy KR, DeLaune RD. Biogeochemistry of Wetlands: Science and Applications. 1st ed. Boca Raton, USA: CRC Press; 2008. 800 p

[7] Rezanezhad F, McCarter CPR, Lennartz B. Editorial: Wetland biogeochemistry: Response to environmental change. Frontiers in Environmental Science [Internet]. 2020;**8**:1-3. Available from: https://www.frontiersin.org/articles/10.3389/fenvs.2020.00055

[8] Döll P, Trautmann T, Göllner M, Schmied HM. A global-scale analysis of water storage dynamics of inland wetlands: Quantifying the impacts of human water use and man-made reservoirs as well as the unavoidable and avoidable impacts of climate change. Ecohydrology [Internet]. 2020;**13**(1):e2175. DOI: 10.1002/eco.2175

[9] Fischer G, Tubiello FN, van Velthuizen H, Wiberg DA. Climate change impacts on irrigation water requirements: Effects of mitigation, 1990-2080. Technological Forecasting and Social Change [Internet]. 2007;**74**(7):1083-1107. Available from: https://www.sciencedirect.com/science/article/pii/S0040162506001429

[10] Siebert S, Henrich V, Frenken K, Burke J. Update of the Digital Global Map of Irrigation Areas to Version 5. Rome, Italy: Food and Agriculture Organization (FAO); 2013

[11] Winkler K, Fuchs R, Rounsevell M, Herold M. Global land use changes are four times greater than previously estimated. Nature Communications [Internet]. 2021;**12**(1):2501. DOI: 10.1038/s41467-021-22702-2

[12] Rosa L, Chiarelli DD, Sangiorgio M, Beltran-Peña AA, Rulli MC, D'Odorico P, et al. Potential for sustainable irrigation expansion in a 3°C warmer climate. Proceedings of the National Academy of Sciences of the United States of America. 2020;**117**(47):29526-29534

[13] Hejazi MI, Edmonds JA, Chaturvedi V. Global irrigation demand - A holistic approach. Irrigation & Drainage Systems Engineering. 2012;**1**(2):1-4

[14] Boretti A, Rosa L. Reassessing the projections of the world water development report. npj Clean Water [Internet]. 2019;**2**(1):15. DOI: 10.1038/s41545-019-0039-9

[15] Rosa L. Adapting agriculture to climate change via sustainable irrigation: Biophysical potentials and feedbacks. Environmental Research Letters. [Internet]. 2022;**17**(6):63008. DOI: 10.1088/1748-9326/ac7408

[16] Rosa L, Chiarelli DD, Rulli MC, Dell'Angelo J, D'Odorico P. Global agricultural economic water scarcity. Science Advances [Internet]. 2020;**6**(18):eaaz6031. DOI: 10.1126/sciadv.aaz6031

[17] Galbraith H, Amerasinghe P, Huber-Lee A. The Effects of Agricultural Irrigation on Wetland Ecosystems in Developing Countries: A Literature Review. (Comprehensive Assessment of Water Management in Agriculture Discussion Paper 1). Colombo, Sri Lanka: International Water Management Institute (IWMI); 2005

[18] Cai X, Zhang X, Noël PH, Shafiee-Jood M. Impacts of climate change on agricultural water management: A review. WIREs Water [Internet]. 2015;**2**(5):439-455. DOI: 10.1002/wat2.1089

[19] Woodhouse P, Veldwisch GJ, Venot JP, Brockington D, Komakech H, Manjichi Â. African farmer-led irrigation development: re-framing agricultural policy and investment? Journal of Peasant Studies [Internet]. 2017;**44**(1):213-233. DOI: 10.1080/03066150.2016.1219719

[20] Higginbottom TP, Adhikari R, Dimova R, Redicker S, Foster T. Performance of large-scale irrigation projects in sub-Saharan Africa. Nature Sustainability [Internet]. 2021;**4**(6):501-508. DOI: 10.1038/s41893-020-00670-7

[21] Kingsford RT, Bino G, Finlayson CM, Falster D, Fitzsimons JA, Gawlik DE, et al. Ramsar wetlands of international importance–improving conservation outcomes. Frontiers in Environmental Science [Internet]. 2021;**9**:1-6. Available from: https://www.frontiersin.org/articles/10.3389/fenvs.2021.643367

[22] Stewart RIA, Dossena M, Bohan DA, Jeppesen E, Kordas RL, Ledger ME, et al. Chapter two - Mesocosm experiments as a tool for ecological climate-change research. In: Woodward G, O'Gorman EJBTA, ER, editors. Global Change in Multispecies Systems: Part 3 [Internet]. Oxford, UK: Academic Press; 2013. pp. 71-181. Available from: https://www.sciencedirect.com/science/article/pii/B9780124171992000021

[23] Walther GR. Community and ecosystem responses to recent climate change. Philosophical Transactions of the Royal Society B: Biological Sciences [Internet]. 2010;**365**(1549):2019-2024. DOI: 10.1098/rstb.2010.0021

[24] Siad SM, Iacobellis V, Zdruli P, Gioia A, Stavi I, Hoogenboom G. A review of coupled hydrologic and crop growth models. Agricultural Water Management [Internet]. 2019;**224**:105746. Available from: https://www.sciencedirect.com/science/article/pii/S0378377419310017

[25] Xia Q, Liu P, Fan Y, Cheng L, An R, Xie K, et al. Representing irrigation processes in the land surface-hydrological

model and a case study in the Yangtze River basin, China. Journal of Advances in Modeling Earth Systems [Internet]. 2022;**14**(7):e2021MS002653. DOI: 10.1029/2021MS002653

[26] Richter BD, Baumgartner JV, Powell J, Braun DP. A method for assessing hydrologic alteration within ecosystems. Conservation Biology [Internet]. 1996;**10**(4):1163-1174. Available from: http://www.jstor.org/stable/2387152

[27] Sheikh V, Sadoddin A, Najafinejad A, Zare A, Hollisaz A, Siroosi H, et al. The density difference and weighted RVA approaches for assessing hydrologic regime alteration. Journal of Hydrology [Internet]. 2022;**613**:128450. Available from: https://www.sciencedirect.com/science/article/pii/S0022169422010204

[28] Gao Y, Vogel RM, Kroll CN, Poff NL, Olden JD. Development of representative indicators of hydrologic alteration. Journal of Hydrology [Internet]. 2009;**374**(1):136-147. Available from: https://www.sciencedirect.com/science/article/pii/S0022169409003321

[29] Yang Z, Yan Y, Liu Q. Assessment of the flow regime alterations in the lower Yellow River, China. Ecological Informatics [Internet]. 2012;**10**:56-64. Available from: https://www.sciencedirect.com/science/article/pii/S157495411100080X

[30] Mathews R, Richter BD. Application of the indicators of hydrologic alteration software in environmental flow setting. JAWRA Journal of the American Water Resources Association [Internet]. 2007;**43**(6):1400-1413. DOI: 10.1111/j.1752-1688.2007.00099.x

[31] Zhou X, Huang X, Zhao H, Ma K. Development of a revised method for indicators of hydrologic alteration for analyzing the cumulative impacts of cascading reservoirs on flow regime. Hydrology and Earth System Sciences [Internet]. 2020;**24**(8):4091-4107. Available from: https://hess.copernicus.org/articles/24/4091/2020/

[32] Richter B, Baumgartner J, Wigington R, Braun D. How much water does a river need? Freshwater Biology [Internet]. 1997;**37**(1):231-249. DOI: 10.1046/j.1365-2427.1997.00153.x

[33] Richter BD, Baumgartner JV, Braun DP, Powell J. A spatial assessment of hydrologic alteration within a river network. Regulated Rivers: Research and Management [Internet]. 1998;**14**(4):329-340. DOI: 10.1002/(SICI)1099-1646(199807/08)14:4%3C329::AID-RRR505%3E3.0.CO

[34] Martínez-Capel F, García-López L, Beyer M. Integrating hydrological modelling and ecosystem functioning for environmental flows in climate change scenarios in the Zambezi River (Zambezi region, Namibia). River Research and Applications [Internet]. 2017;**33**(2):258-275. DOI: 10.1002/rra.3058

[35] Hillman B, Douglas EM, Terkla D. An analysis of the allocation of Yakima River water in terms of sustainability and economic efficiency. Journal of Environmental Management [Internet]. 2012;**103**:102-112. Available from: https://www.sciencedirect.com/science/article/pii/S0301479712000795

[36] Stefanidis K, Panagopoulos Y, Psomas A, Mimikou M. Assessment of the natural flow regime in a Mediterranean river impacted from irrigated agriculture. Science of the Total Environment. 2016;**573**:1492-1502

[37] Xue L, Zhang H, Yang C, Zhang L, Sun C. Quantitative assessment of hydrological alteration caused

by irrigation projects in the Tarim River basin, China. Scientific Reports [Internet]. 2017;**7**(1):4291. DOI: 10.1038/s41598-017-04583-y

[38] Yasarer LMW, Taylor JM, Rigby JR, Locke MA. Trends in land use, irrigation, and streamflow alteration in the Mississippi River alluvial plain. Frontiers in Environmental Science [Internet]. 2020;**8**:1-13. Available from: https://www.frontiersin.org/articles/10.3389/fenvs.2020.00066

[39] Wang X, Ma W, Lv J, Li H, Liu H, Mu G, et al. Analysis of changes in the hydrological regime in Lalin River basin and its impact on the ecological environment. Frontiers in Earth Science [Internet]. 2022;**10**:1-16. Available from: https://www.frontiersin.org/articles/10.3389/feart.2022.987296

[40] Barungi M, Odokonyero T. Understanding the rice value chain in Uganda: Opportunities and challenges to increased productivity. [Internet]. In: Research Report No. 15. Kampala, Uganda: Economic Policy Research Centre; 2016. Available from: https://eprcug.org/publication/understanding-the-rice-value-chain-in-uganda-opportunities-and-challenges-to-increased-production/?wpdmdl=11868&refresh=6426c9b0647e11680263600&ind=1603370391868&filename=UnderstandingthericevaluechaininUganda-opport

[41] Bunyangha J, Majaliwa MJG, Muthumbi AW, Gichuki NN, Egeru A. Past and future land use/land cover changes from multi-temporal landsat imagery in Mpologoma catchment, eastern Uganda. The Egyptian Journal of Remote Sensing and Space Science [Internet]. 2021;**24**(3, Part 2):675-685. Available from: https://www.sciencedirect.com/science/article/pii/S1110982321000181

[42] Chombo O, Lwasa S, Makooma TM. Spatial differentiation of small holder farmers' vulnerability to climate change in the Kyoga Plains of Uganda. American Journal of Climate Change. 2018;**7**(04):624

[43] Namutebi J, Sekanjako H. Food security: Govt to borrow sh187b. New Vision [Internet]. 2013:32. Available from: https://www.newvision.co.ug/new_vision/news/1314210/food-security-govt-borrow-sh187b

[44] MAAIF. Enhancing National Food Security through Increased Rice Production Project (ENRP) [Internet]. Kampala, Uganda: The ENRP Project; 2019. Available from: https://www.agriculture.go.ug/enhancing-national-food-security-through-increased-rice-production-project/ [Accessed: November 26, 2019]

[45] MAAIF. Detailed studies, detailed engineering design and construction supervision for proposed irrigation schemes development of Igogero and Naigombwa in Bugiri and Bugweri districts. In: Final Detailed Design Report - Project Works Phase 1. Kampala, Uganda: Ministry of Agriculture, Animal Industry, and Fisheries (MAAIF) of Uganda; 2021

[46] MWE. Mpologoma Catchment Management Plan. Kampala, Uganda: Ministry of Water and Environment (MWE) of Uganda; 2018

[47] Semeniuk CA, Semeniuk V. A geomorphic approach to global classification for inland wetlands. Vegetatio [Internet]. 1995;**118**(1):103-124. DOI: 10.1007/BF00045193

[48] Kipkemboi J, Van Dam AA. Papyrus Wetlands. In: Finlayson C, Milton G, Prentice R, Davidson N, editors. The Wetland Book. Springer: Dordrecht; 2018. pp. 183-197

[49] Chapman L, Balirwa J, Bugenyi FWB, Chapman C, Crisman T. Wetlands of East Africa: Biodiversity, exploitation and policy perspectives. In: Gopal B, Junk WJ, Davis JA, editors. Biodiversity in Wetlands: Assessment, Function and Conservation. Vol. 2. Leiden, Netherlands: Backhuys Publishers; 2001. pp. 101-131

[50] Ewen J, Geoff P, O'Connell EP. Shetran: Distributed River Basin flow and transport Modeling system. Journal of Hydrologic Engineering [Internet]. 2000;**5**(3):250-258. DOI: 10.1061/(ASCE)1084-0699(2000)5:3(250)

[51] Birkinshaw SJ, James P, Ewen J. Graphical user interface for rapid set-up of SHETRAN physically-based river catchment model. Environmental Modelling & Software [Internet]. 2010;**25**(4):609-610. Available from: https://www.sciencedirect.com/science/article/pii/S1364815209003004

[52] Birkinshaw SJ. Technical note: Automatic river network generation for a physically-based river catchment model. Hydrology and Earth System Sciences Discussions [Internet]. 2010;**14**(9):1767-1771. Available from: https://hess.copernicus.org/articles/14/1767/2010/

[53] Oyarmoi A, Birkinshaw S, Hewett CJM, Fowler HJ. The effect of papyrus wetlands on flow regulation in a Tropical River catchment. Land [Internet]. 2023;**12**(12):2158. Available from: https://www.mdpi.com/2073-445X/12/12/2158

[54] Rangecroft S, Birkinshaw S, Rohse M, Day R, McEwen L, Makaya E, et al. Hydrological modelling as a tool for interdisciplinary workshops on future drought. Progress in Physical Geography [Internet]. 2018;**42**(2):237-256. DOI: 10.1177/0309133318766802

[55] Birkinshaw S. SHETRAN Hydrological model [Internet]. Newcastle, UK: Newcastle University; 2021. Available from: https://research.ncl.ac.uk/shetran/index.htm [Accessed: October 4, 2024]

[56] Orasen G, De Nisi P, Lucchini G, Abruzzese A, Pesenti M, Maghrebi M, et al. Continuous flooding or alternate wetting and drying differently affect the accumulation of health-promoting phytochemicals and minerals in rice brown grain. Agronomy. 2019;**9**(10):628

[57] Carrijo DR, Lundy ME, Linquist BA. Rice yields and water use under alternate wetting and drying irrigation: A meta-analysis. Field Crops Research [Internet]. 2017;**203**:173-180. Available from: https://www.sciencedirect.com/science/article/pii/S0378429016307791

[58] IRRI. Rice farming: Saving water through alternate wetting drying (AWD) method, Indonesia. 2013. Available from: https://www.fao.org/3/ca4023en/ca4023en.pdf

[59] Bin D, Zhichen L, Loeve R, Molden D, Baozhong Y. Rice impact in Henan irrigation districts along the lower Yellow River reaches. In: Yellow River Conservancy Commission. In: First International Yellow River Forum on River Basin Management [Internet]. Zhengzhou, China: The Yellow River Conservancy Publishing House; 2003. pp. 105-113. Available from: https://www.iwmi.cgiar.org/assessment/files_new/publications/WorkshopPapers/IYRF_2003_Dong-Bin.pdf

[60] Brouwer C, Heibloem M. Irrigation Water Management. Training Manual No. 3 - Irrigation Water Needs [Internet]. Rome, Italy: Food and Agriculture Organization (FAO); 1986. Available from: http://www.fao.org/docrep/S2022E/S2022E00.htm

[61] FAO. AquaCrop. In: The Crop Water Productivity Model [Internet]. Rome, Italy: Land and Water Division of FAO; 2009. Available from: https://www.fao.org/aquacrop/en/

[62] Raes D, Steduto P, Hsiao TC, Fereres E. AquaCrop Version 7.0 Reference Manual [Internet]. Rome, Italy: Food and Agriculture Organization (FAO); 2022. Available from: https://www.fao.org/aquacrop/resources/referencemanuals/en/

[63] Raes D. AquaCrop Training Handbooks. Book I - Understanding AquaCrop [Internet]. Rome, Italy: Food and Agriculture Organization (FAO); 2022. Available from: https://www.fao.org/3/cc2380en/cc2380en.pdf

[64] IRRI. Planting the Rice: Nursery Systems. Los Baños, Philippines: International Rice Research Institute (IRRI) [Internet]; 2007. Available from: http://www.knowledgebank.irri.org/ericeproduction/II.5_Nursery_systems.htm [Accessed: Feburary 15, 2023]

[65] Guerra LC, Bhuiyan SI, Tuong TP, Barker R. Producing More Rice with Less Water from Irrigated System. Manila: Philippines; 1998

[66] Richards MB, Sander BO. Alternate Wetting and Drying in Irrigated Rice. Climate-Smart Agriculture Practice Brief [Internet]. Copenhagen, Denmark: CGIAR Research Program on Climate Change, Agriculture and Food Security; 2014. Available from: https://cgspace.cgiar.org/handle/10568/35402

[67] Vila-Traver J, González de Molina M, Infante-Amate J, Aguilera E. Disentangling the effect of climate and cropland changes on the water performance of agroecosystems (Spain, 1922-2016). Journal of Cleaner Production [Internet]. 2022;**344**:130811. Available from: https://www.sciencedirect.com/science/article/pii/S0959652622004498

[68] The Nature Conservancy. Indicators of Hydrologic Alteration. Version 7.1 User's Manual. Virginia, USA: The Nature Conservancy (TNC); 2009. p. 81

[69] Helsel DR, Hirsch RM. Statistical methods in water resources. In: Statistical Methods in Water Resources: Techniques of Water-Resources Investigations of the United States Geological Survey Book 4, Hydrologic Analysis and Interpretation. Version 1. [Internet]. Reston, VA: U.S. Geological Survey; 2002. p. 524. Available from: http://pubs.er.usgs.gov/publication/twri04A3

[70] Essaid HI, Caldwell RR. Evaluating the impact of irrigation on surface water - groundwater interaction and stream temperature in an agricultural watershed. Science of the Total Environment. 2017;**599-600**:581-596

[71] Negri C, Chiaradia E, Rienzner M, Mayer A, Gandolfi C, Romani M, et al. On the effects of winter flooding on the hydrological balance of rice areas in northern Italy. Journal of Hydrology. 2020;**590**:125401. Available from: https://www.sciencedirect.com/science/article/pii/S0022169420308611

[72] Wang H, Xu S, Sun L. Effects of climatic change on evapotranspiration in Zhalong wetland, Northeast China. Chinese Geographical Science [Internet]. 2006;**16**(3):265-269. DOI: 10.1007/s11769-006-0265-1

[73] Li Z, Ciais P, Wright JS, Wang Y, Liu S, Wang J, et al. Increased precipitation over land due to climate feedback of large-scale bioenergy cultivation. Nature Communications [Internet]. 2023;**14**(1):4096. DOI: 10.1038/s41467-023-39803-9

[74] Makula EK, Zhou B. Coupled model Intercomparison project phase 6 evaluation and projection of east African precipitation. International Journal of Climatology [Internet]. 2022;**42**(4):2398-2412. DOI: 10.1002/joc.7373

[75] Ayugi B, Shilenje ZW, Babaousmail H, Lim Kam Sian KTC, Mumo R, Dike VN, et al. Projected changes in meteorological drought over East Africa inferred from bias-adjusted CMIP6 models. Natural Hazards [Internet]. 2022;**113**(2):1151-1176. DOI: 10.1007/s11069-022-05341-8

[76] Almazroui M, Saeed F, Saeed S, Nazrul Islam M, Ismail M, Klutse NAB, et al. Projected change in temperature and precipitation over Africa from CMIP6. Earth Systems and Environment [Internet]. 2020;**4**(3):455-475. DOI: 10.1007/s41748-020-00161-x

[77] Olden JD, Poff NL. Redundancy and the choice of hydrologic indices for characterizing streamflow regimes. River Research and Applications [Internet]. 2003;**19**(2):101-121. DOI: 10.1002/rra.700

[78] Cui Z, Wang Y, Zhang GJ, Yang M, Liu J, Wei L. Effects of improved simulation of precipitation on evapotranspiration and its partitioning over land. Geophysical Research Letters [Internet]. 2022;**49**(5):e2021GL097353. DOI: 10.1029/2021GL097353

[79] Ayugi B, Dike V, Ngoma H, Babaousmail H, Mumo R, Ongoma V. Future changes in precipitation extremes over East Africa based on CMIP6 models. Water. 2021;**2358**:1-19

[80] Shivanna KR. Climate change and its impact on biodiversity and human welfare. Proceedings of the Indian National Science Academy [Internet]. 2022;**88**(2):160-171. DOI: 10.1007/s43538-022-00073-6

[81] IPCC. Summary for policymakers. In: Masson-Delmotte V, Zhai P, Pörtner HO, Roberts D, Skea J, Shukla PR, et al., editors. Global Warming of 15°C An IPCC Special Report on the Impacts of Global Warming of 15°C above Pre-Industrial Levels and Related Global Greenhouse Gas Emission Pathways, in the Context of Strengthening the Global Response to the Threat of Climate Change, [Internet]. Cambridge, UK and New York, NY, USA: Cambridge University Press; 2018. pp. 1-24. Available from: https://www.cambridge.org/core/product/identifier/9781009157940%23prf2/type/book_part

[82] Ramalho Q, Vale MM, Manes S, Diniz P, Malecha A, Prevedello JA. Evidence of stronger range shift response to ongoing climate change by ectotherms and high-latitude species. Biological Conservation [Internet]. 2023;**279**:109911. Available from: https://www.sciencedirect.com/science/article/pii/S0006320723000113

[83] Couet J, Marjakangas EL, Santangeli A, Kålås JA, Lindström Å, Lehikoinen A. Short-lived species move uphill faster under climate change. Oecologia [Internet]. 2022;**198**(4):877-888. DOI: 10.1007/s00442-021-05094-4

[84] Velásquez-Tibatá J, Salaman P, Graham CH. Effects of climate change on species distribution, community structure, and conservation of birds in protected areas in Colombia. Regional Environmental Change [Internet]. 2013;**13**(2):235-248. DOI: 10.1007/s10113-012-0329-y

[85] Pompeu CR, Peñas FJ, Goldenberg-Vilar A, Álvarez-Cabria M, Barquín J. Assessing the effects of

irrigation and hydropower dams on river communities using taxonomic and multiple trait-based approaches. Ecological Indicators [Internet]. 2022;**145**:109662. Available from: https://www.sciencedirect.com/science/article/pii/S1470160X22011359

[86] Guengant JP, May JF. Africa 2050. In: African Demography [Internet]. Abidjan, Côte d'Ivoire: Emerging Markets Forum (EMF); 2013. Available from: https://horizon.documentation.ird.fr/exl-doc/pleins_textes/divers13-07/010059333.pdf

[87] Kashaigili JJ, McCartney MP, Mahoo HF, Lankford BA, Mbilinyi BP, Yawson DK. Use of a hydrological model for environmental Management of the Usangu Wetlands, Tanzania. In: IWMI Research Report 104. Colombo, Sri Lanka: International Water Management Institute (IWMI); 2009

[88] Haileselassie H, Araya A, Habtu S, Meles KG, Gebru G, Kisekka I, et al. Exploring optimal farm resources management strategy for Quncho-teff (Eragrostis tef (Zucc.) trotter) using AquaCrop model. Agricultural Water Management [Internet]. 2016;**178**:148-158. Available from: https://www.sciencedirect.com/science/article/pii/S0378377416303328

[89] Singh SK, Ibbitt R, Srinivasan MS, Shankar U. Inter-comparison of experimental catchment data and hydrological modelling. Journal of Hydrology [Internet]. 2017;**550**:1-11. Available from: https://www.sciencedirect.com/science/article/pii/S002216941730269X

[90] Seibert J, McDonnell JJ. On the dialog between experimentalist and modeler in catchment hydrology: Use of soft data for multicriteria model calibration. Water Resources Research [Internet]. 2002;**38**(11):14-23. DOI: 10.1029/2001WR000978

[91] Cerdan O, Le Bissonnais Y, Govers G, Lecomte V, van Oost K, Couturier A, et al. Scale effect on runoff from experimental plots to catchments in agricultural areas in Normandy. Journal of Hydrology [Internet]. 2004;**299**(1):4-14. Available from: https://www.sciencedirect.com/science/article/pii/S0022169404001118

[92] Saxton KE, Rawls W. Soil Water Characteristics - Hydraulic Properties Calculator [Internet]. USDA Agricultural Research Service and Department of Biological Systems Engineering: Washington State University; 2018. Available from: https://hrsl.ba.ars.usda.gov/soilwater/Index.htm

[93] Dembélé M, Schaefli B, van de Giesen N, Mariéthoz G. Suitability of 17 gridded rainfall and temperature datasets for large-scale hydrological modelling in West Africa. Hydrology and Earth System Sciences [Internet]. 2020;**24**(11):5379-5406. Available from: https://hess.copernicus.org/articles/24/5379/2020/

[94] Bisselink B, Zambrano-Bigiarini M, Burek P, de Roo A. Assessing the role of uncertain precipitation estimates on the robustness of hydrological model parameters under highly variable climate conditions. Journal of Hydrology: Regional Studies [Internet]. 2016;**8**:112-129. Available from: http://www.sciencedirect.com/science/article/pii/S2214581816300817

[95] Prudhomme C, Jakob D, Svensson C. Uncertainty and climate change impact on the flood regime of small UK catchments. Journal of Hydrology [Internet]. 2003;**277**(1):1-23. Available

from: https://www.sciencedirect.com/science/article/pii/S0022169403000659

[96] Lee S, Qi J, McCarty GW, Yeo IY, Zhang X, Moglen GE, et al. Uncertainty assessment of multi-parameter, multi-GCM, and multi-RCP simulations for streamflow and non-floodplain wetland (NFW) water storage. Journal of Hydrology. 2021;**600**:126564

Chapter 2

Innovative Irrigation System to Enhance the Water Productivity in Closed Spaced Crops

Vallal Kannan Sankaralingam, Marimuthu Subramanian, Sangeetha Jebalin Vijayalalitha Vallal Kannan and HariniHarishma Veerappan Prithviraj

Abstract

Water plays a crucial role in supporting plant growth and development from seed germination to maturity ultimately contributing to the expected yield. However, the availability and quality of water for irrigation are declining, while its diversion for other uses continues to rise annually. Advanced irrigation systems enhance water use efficiency. The adoption and sustainability of these systems are closely linked to the users' expertise in operation and ongoing upkeep. Rain hose irrigation systems, known for their minimal maintenance requirements and ease of operation, are gaining popularity as an advanced alternative for closely spaced crops. These systems provide high water use efficiency and productivity with less need for technical knowledge, making them highly sustainable and accessible. Studies on closely spaced crops like blackgram revealed that rain hose systems maintained the highest soil moisture levels (23.93% at 10 cm depth and 19.71% at 20 cm depth). The rain hose system proved more cost-effective, reducing costs by 27.09% and achieving a 15.23% higher benefit-cost ratio. Similarly, in groundnut research, the rain hose system at 100% PET delivered the highest benefit-cost ratio (3.29) due to its lower installation costs. It also performed comparably well across other parameters, including pod yield, lower disease incidence, higher hundred-pod weight, increased oil yield, and shelling percentage. The rain hose irrigation system is farmer-friendly, with low establishment costs, minimal skill requirements, and suitability for small and marginal farmers making it a viable and efficient irrigation option.

Keywords: drip irrigation, rain hose irrigation, sprinkler irrigation, water productivity, water use efficiency

1. Introduction

In Asia, approximately 84% of water withdrawals are allocated to agriculture, whereas the global average stands at 71%. Water is the single most important component for sustainable agricultural production, especially in the traditional method of irrigation practices. In World Water Demand and Supply, 1990 to 2025, the IWMI forecasts an increase in water demand for irrigation under two scenarios. In the first scenario, irrigation efficiency remains at its 1990 level until 2025, while in the second scenario, efficiency improves over time. The assumptions are that the per capita amount of food production from irrigated agriculture will remain constant.

The irrigation potential of India is 113.5 million hectares, corresponds to the gross area which could theoretically be irrigated in a year on the basis of the assumed design cropping pattern and a rainfall probability of 75%, and represents 2.27 times the area under irrigation in 1993. The figure of 2000 m^3/inhabitant/year is usually used as an indicator of water scarcity and India reached this limit with the deficiency ratio of 33.9%. Producing more crops with less water is therefore a formidable challenge for the food, economic, social, and water security of the nation.

India's total irrigated area stands at 112,229,000 hectares, with a net irrigated area of 754,560,000 hectares [1]. This signifies that 55% of India's agricultural land now has access to irrigation marking an increase from 47.8% in 2013–2014. In the fiscal year 2022–2023, 73 million hectares, equivalent to 52% of India's cultivated land, had access to irrigation. In Tamil Nadu, the irrigated area spans 2.15 million hectares, constituting 54% of the state's total cropped area.

Currently, irrigation systems waste approximately 50% of applied water. Drip irrigation stands out as the most efficient method, wasting only around 10% of water. Spray irrigation, on the other hand, squanders roughly 35% of applied water, but this can be reduced to about 10% through the adoption of overhead pipes. There is a pressing need for irrigation systems with higher water use efficiency that are farmer-friendly, ensuring ease of operation and maintenance while maximizing water productivity. Moreover, water-saving practices, which demand enhanced water control, are often integrated into broader agronomic improvement strategies aimed at optimizing input efficiency. As a result, they play a crucial role in enhancing total factor productivity. Consequently, irrigation systems contribute not only to increasing water use or irrigation efficiency but also to sustaining or improving overall water productivity.

2. Need of water for 2025

As per the findings of the National Commission on Integrated Water Resources Development, India's irrigation requirements are projected to reach 611 billion cubic meters (BCM) by 2025. This estimate is derived from an average weighted value of 0.73 meters for surface water and 0.51 meters for groundwater. The irrigation sector in India is anticipated to continue being the largest consumer of water, with usage expected to increase from 688 BCM in 2010 to 910 BCM in 2025 and further to 1072 BCM by 2050.

2.1 Irrigations systems in agriculture

Traditional agricultural methods involve manual labor and animal assistance, employing techniques such as moats, chain-pumps, dhekli, and rahat. In contrast, modern irrigation methods encompass technologies like sprinkler systems and drip irrigation.

2.2 Modern methods of irrigation

Modern irrigation methods offer solutions to the shortcomings of traditional techniques, promoting efficient water usage. These methods primarily include the sprinkler and drip irrigation systems.

Sprinkler System: Water is dispersed over the crops either from a central location *via* high-pressure overhead sprinklers or from moving platforms equipped with sprinklers. This method ensures uniform water distribution and is particularly beneficial in areas experiencing water scarcity. A pump connected to pipes generates pressure, leading to water being sprinkled through pipe nozzles. Sprinkler systems typically exhibit a water use efficiency of 80–85%, although actual efficiency may range between 60 and 90%, averaging at around 75%. Centre-pivot systems typically achieve an efficiency of approximately 85%.

Drip System: The drip irrigation system delivers water directly to the roots of plants, drop by drop, *via* hoses or pipes. It is suitable for regions with limited water availability. This method minimizes water wastage and runoff. While drip irrigation demands more maintenance, it boasts impressive water use efficiency, ranging between 95 and 100%. Additionally, it facilitates efficient plant water utilization, with an efficiency rate of around 90%. By delivering water precisely to the plant root zone, drip irrigation reduces evaporation and runoff.

2.3 Constraints and limitations in modern irrigation systems

Drip irrigation stands out as the most water-efficient method for irrigating various types of plantings. Particularly well-suited for clay soils, this method applies water slowly, allowing the soil to absorb it effectively and prevent runoff. Compared to overhead spray devices, drip devices utilize only a fraction of the water. However, despite its numerous advantages, drip irrigation systems come with several drawbacks to consider:

High initial cost: The upfront investment for installing a drip irrigation system can be relatively steep compared to traditional methods. This includes expenses for materials like emitters, filters, valves, and the installation of system components.

System complexity: Designing and installing drip irrigation systems can be complex and require technical expertise. Proper planning and layout are crucial to ensure uniform water distribution and prevent issues like clogging or emitter malfunction.

Maintenance requirements: Regular maintenance is necessary to keep drip irrigation systems functioning optimally. Emitters may become clogged with debris or mineral deposits, necessitating frequent cleaning or replacement. Filters may also require periodic cleaning or replacement.

Vulnerability to damage: Drip irrigation components, including tubes and emitters, are prone to damage from rodents, insects, or accidental human activity. Such physical damage can cause leaks or disrupt water flow, reducing system efficiency.

Challenges in system design: Properly designing a drip irrigation system requires careful consideration of factors like soil type, topography, plant spacing, and water pressure. Poor design or layout can lead to uneven water distribution and inadequate coverage.

Limited water flow: Operating at relatively low pressure, drip irrigation systems may deliver less water compared to other methods. This limitation can be problematic during peak water demand or for crops requiring high water inputs.

Risk of clogging: Drip systems are susceptible to clogging, particularly when the water source contains sediments, debris, or high mineral content. Blocked emitters can disrupt water distribution, lower efficiency, and necessitate frequent maintenance.

Alternate method of irrigation paves the way to solve the limitations and constraints of modern irrigations systems. With this back ground, rain hose irrigation was used to assess the suitability for closed crops and also supports to overcome the constraints of modern irrigation systems.

2.4 Alternate method with high use efficiency and low-cost irrigation system

Rain hose irrigation, also known as rain pipes or micro irrigation spray tubes, uses perforated tubes with laser-punched holes to deliver water efficiently. These cost-effective systems offer over 60% water savings, ensure uniform distribution, and target plant roots to minimize wastage. Lightweight and portable, they are easy to install and maintain.

2.5 Suitability of low-cost irrigation for agricultural crops

Rain hose irrigation systems are well-suited for densely planted crops such as onions, various vegetables, leafy greens, pulses, and groundnuts. These systems, also known as rain pipe irrigation, employ a flexible hose featuring a series of drip holes designed to spray water up to 10–15 feet on either side. Utilizing nano punching technology, these drip holes ensure a consistent water flow. Rain hose irrigation boasts the capability to conserve over 60% of water while irrigating crops at an accelerated pace, leading to savings in electricity and labor. The utilization of nano punching technology ensures uniform water distribution, with the laser-punched holes less prone to clogging.

Research and investment efforts have been directed towards implementing cost-effective technology, precision agriculture, and environmentally friendly practices to promote sustainable water usage in agricultural development. Rain hose technology exemplifies an affordable spray irrigation solution, serving as a substitute for traditional sprinkler systems. It offers easy installation and maintenance, featuring a flexible hose with a series of drip holes crafted using nano punching technology to ensure consistent water distribution. It represents an emerging irrigation technique widely adopted for densely planted crops, functioning on a spraying mechanism akin to sprinkler irrigation. The spraying pattern of rain hose irrigation typically extends linearly on both sides along the full length of the hose (up to 50 meters). Studies have investigated variations in spray width and water discharge from nozzles across different flow rates and rain hose diameters. However, there is currently no standardized recommendation regarding the optimal combination of rain hose length and diameter. Field tests employing the Catch-can method have been conducted to evaluate the influence of rain hose length and diameter across various flow rates and pressure values within the main pipe.

Rain pipe irrigation has emerged as a promising solution for maximizing water utilization and enhancing crop yields at a reduced cost compared to traditional irrigation systems. An experiment was conducted to assess the hydraulic performance of rain pipes of varying lengths. The findings revealed that the optimal performance of the rain pipe system was achieved when operated at a pressure of 1.50 kg/cm^2, with a rain pipe length of 30 meters and a spacing of 4 meters. This configuration resulted in the highest uniformity coefficient (87.83%), distribution uniformity (76.29%), and mean application rate (6.81 cm/h). Moreover, utilizing a 30-meter length at 1.5 kg/cm^2 led to the maximum discharge per meter length of rain pipe and the widest coverage width. These results underscore the effectiveness of rain pipe irrigation, particularly when employing an operating pressure of 1.5 kg/cm^2, a rain pipe length of 30 meters, and a spacing of 4 meters. The study underscores the suitability of rain pipe irrigation as a cost-effective and efficient alternative for water management in

Indian agriculture. By optimizing factors such as operating pressure, rain pipe length, and spacing, farmers can enhance water use efficiency and bolster crop productivity, thereby contributing to sustainable agricultural practices amidst resource constraints.

3. Case study

3.1 Low-cost irrigation technology for pulse productivity and profitability

Most pulses contribute significantly to nutritional security for humans while also providing environmental benefits by improving soil health. Cultivated over an area of approximately 95.75 million hectares, their average productivity is about 1.0 t/ha. Pulses tend to perform better under irrigated conditions, as irrigation plays a critical role in enhancing their productivity. The water requirements for pulse crops vary significantly, ranging from as low as 5.7 mm during the rainy season to 609 mm in the dry season. Consequently, the average water productivity of major pulses ranges between 0.67 and 7.51 kg/ha mm. Innovative strategies to improve water productivity can help reduce water losses and boost yields. For example, adopting micro-irrigation systems for pulses significantly reduces water usage while increasing yields. In water-scarce conditions, deficit irrigation or irrigation based on critical growth stages can also enhance water productivity. Techniques such as bed planting, and ridge-and-furrow planting are proven to increase water productivity by 20–40% due to improved aeration, reduced lodging, and easier intercultural operations. Furthermore, irrigation methods strongly influence the quality of pulses. Pulse production demonstrates a linear correlation with irrigation water as a key input during the growing season.

The value of water productivity, which is one of the yardsticks for measuring how efficiently pulse crops use water, needs to be enhanced by suitable means. Water productivity is a simple ratio of crop yield to total amount of water used by the crop during its growing period and can be improved by increasing output with the same amount of water or by producing the same output with less water [2]. Black gram cultivated in the summer season requires more water compared to its cultivation in the winter, with water usage ranging between 400 and 500 mm. Sustainable approaches to improving crop water productivity are becoming increasingly vital in areas where water is the primary limitation to agricultural production. Black gram, a significant pulse crop, is cultivated on 41% of the total pulse-growing area in Tamil Nadu, covering approximately 3.40 lakh hectares with a production of 1.20 lakh tons, resulting in an average productivity of 328 kg/ha.

3.1.1 Materials and methods

The experiment was conducted at Agricultural Engineering College and Research Institute (Tamil Nadu Agricultural University) in Kumulur, Tiruchirappalli, Tamil Nadu, India. The climatic conditions included a total rainfall of 98 mm, with 50% being effective, spread over six rainy days. The average pan evaporation rate was 6.12 mm per day. The experiments were carried out following the recommended cultivation practices for black gram, specifically focusing on the VBN11 variety which has growth duration of 70–75 days. The treatments were organized using a Factorial Randomized Block Design (RBD) with four replications. The experiment consists of five treatments of Check Basin Method (T_1), Raised Bed Method (T_2), Drip Irrigation System (T_3) Sprinkler Irrigation System (T_4), and Rain Hose Irrigation System (T_5).

An area of 0.04 hectares was designated for each irrigation method incorporating various land configurations as outlined.

A uniform application of 25:50:25 kg/ha of NPK (Nitrogen, Phosphorus, Potassium) was implemented across all irrigation methods as a standard practice. The standard agronomic practices were followed on cultivation of black gram. Irrigation was delivered using a drip system with the initial watering carried out immediately after sowing. Subsequent irrigations were scheduled every 3 days, based on 100% of the daily pan evapo-transpiration (PET). In the rain hose irrigation method, the system was utilized on flat bed configurations, whereas check basin irrigation was applied to check basin land configurations. The irrigation frequency for these systems was adjusted to once every 7 days determined by 100% PET calculations using weather data from each location.

Growth and yield parameters, water indices and economics were observed and calculated and presented in tables. The economic analysis of each treatment was conducted by accounting for all aspects of crop production and the current market prices of inputs. For irrigation methods, the evaluation included installation and field preparation costs along with considerations for their lifespan and depreciation. For example, the lifespan of drip and sprinkler irrigation systems was assumed to be 7 years, while that of rain hose irrigation was estimated at 5 years. Gross returns were determined by multiplying the grain yield by the prevailing market price.

3.1.2 Results and discussion

3.1.2.1 Effect of treatments on yield and economics

The results indicated that among the various irrigation methods, drip fertigation recorded the significantly highest seed yield of 1363 kg/ha recorded in drip irrigation; the rain hose method of irrigation system was performed next in order; however, it was found to be statistically equivalent with the seed yield of 1172 kg/ha as indicated in **Table 1**. In contrast, the check basin method of irrigation (T_1) recorded the lowest values of yield.

3.1.2.2 Effect of treatment on water indices

Crop productivity per unit of water used was evaluated through an analysis of water use efficiency and water productivity. A total of 280.3 mm of water, including

Treatment	Seed yield (kg/ha)	Gross return (Rs/ha)	Net return (Rs/ha)	BCR
T_1	825	49,485	22,690	1.86
T_2	964	57,810	29,641	2.07
T_3	1363	81,780	44,460	2.10
T_4	1063	64,020	36,081	2.23
T_5	1172	70,890	43,220	2.48
Sed	12.97	—	—	—
CD (p = 0.05)	28.27	—	—	—

Table 1.
Effect of treatments on yield and economics.

rainfall during the cropping period, was applied to the black gram crop using various irrigation systems. The findings showed that although all methods used the same amount of water, drip irrigation achieved the highest average water use efficiency of 48.6 kg/ha mm and the highest water productivity of Rs. 292 ha/mm. However, the check basin irrigation method (T_1) recorded the lowest water use efficiency and water productivity [3].

In summary, adopting rain hose irrigation can significantly enhance growth, productivity, soil moisture retention, water use efficiency, and overall profitability for farmers. While drip irrigation offers superior yields and nutrient uptake, the rain hose system provides distinct advantages, including reduced costs and prolonged soil moisture availability. Cultivating black gram with rain hose irrigation reduces costs by 27.1% and achieves a 15.32% higher benefit-cost ratio (BCR) compared to drip irrigation. This approach ensures efficient nutrient solubilization and delivery to the root zone, fostering favorable microclimates for better black gram growth and yield. Additionally, incorporating fertigation with water-soluble fertilizers through drip irrigation improves nutrient use efficiency and reduces environmental impacts. However, the high cost of water-soluble fertilizers poses a challenge for some farmers. Ultimately, selecting an appropriate irrigation method, optimizing nutrient management, and prioritizing cost-effective practices are essential for sustainable and profitable black gram cultivation.

3.2 Low-cost water management techniques in groundnut for higher productivity and profitability

The water required by groundnut to complete its life cycle is around 400–600 mm. Among all the agronomic practices, water management stands second that contributes yield next to fertilizer application of around 27%. Optimum soil moisture condition needs to maintain by irrigating the crop with adequate moisture content for optimum plant population and higher yield. The higher moisture levels will encourage high vegetative growth as well as inhibit the flower initiation and development, and consequently, productivity will be reduced. The adequate soil moisture level during the vegetative growth phase leads to better flowering initiation thus enhancing the pod yield will give higher productivity. By considering all those parameters, finding out the best irrigation management approach that provides optimal soil moisture needs to be cost-effective and must influence higher productivity and higher profitability.

In comparison with traditional irrigation practices, micro-irrigation systems play a crucial role in saving agricultural water use. Micro-irrigation is an appropriate and efficient way to provide water closer to the plant and directly to the soil with less water loss, increasing water productivity [4]. Adopting suitable irrigation management technologies related to the specific crop, significantly influence on water use efficiency and improve productivity in a cost-effective manner. The irrigation management technologies maximize productivity by providing adequate moisture to the crop during the crop growth stages [5].

Deficit irrigation is currently considered a significant and sustainable production strategy that maximizes water productivity by limiting the amount of water applied to plants [6]. Deficit irrigation was said to enhance water productivity while saving a significant amount of irrigation water for various crops, especially in dry and semi-arid areas [7].

In dry regions, deficit irrigation has been employed to maximize crop water efficiency on-farm strategy. With this backdrop, the research has been conducted at

Agricultural College and Research Institute, Madurai, with the objectives to find out the optimum moisture for low vegetative growth and high yield by profuse flowering and pod setting and to find out the best water management approach for higher productivity and higher profitability.

3.2.1 Materials and methods

The crop was cultivated during *rabi* season (January to May) of 2021 with VRI 8 (Virudhachalam 8) groundnut variety having the duration of 105–110 days to reach full maturity. The main field experiment was laid out in a strip plot design with three replications.

The field experiment consists of 12 treatment combinations with split plot design. The main plot treatment includes four different methods of irrigation system: drip irrigation (I_1), sprinkler irrigation (I_2), rain hose irrigation (I_3), and broad bed and furrow irrigation (I_4), and the subplot treatment includes three levels of moisture stress conditions *viz.*, 100% of Potential Evapotranspiration (PET) (S_1), 125% of PET (S_2) and 75% of PET(S_3).

The irrigation was given to the crop once in 3 days intervals, and the quantity of water has been decided based on the pan evaporation method.

$$ET_O = E_{pan} \times K_p \quad (1)$$

Where ET_O—potential evapotranspiration; E_{pan}—pan evaporation (mm); K_p—pan coefficient (0.8)

ET_O is the potential evapotranspiration; it was calculated by using the Class A pan evaporimeter.

The quantity of water that needs to be given was calculated at different levels of 100%, 125%, and 75% of PET by using the pan evaporation method, which was given to the crop through the drip, sprinkler, rain hose, and BBF method of irrigation systems. The water (mm) evaporated in the 3 days interval had been calculated and supplemented by using the pan evaporation method. The specifications for each method of irrigation were listed in **Table 2.**

Irrigation method	Discharge rate (l/h)	Pressure (kg/cm^2)	Specifications
Drip	4 l/h (per emitter) 8 l/h/m^2	1.5–2	Bed size—90 cm Furrow size—30 cm Lateral spacing—120 cm Lateral size—16 mm Emitter spacing—40 cm
Sprinkler	1920 l/h 13.3 l/h/m^2	2–3	Wetting diameter—12 m
Rain hose	200 l/h/m of pipe 33.3 l/h/m^2	1	Wetting diameter—6 m Pipe diameter—40 mm
BBF	—	—	Bed size—120 cm Furrow—30 cm

Table 2.
Discharge rate, pressure along with some specifications of different methods of irrigation.

Irrigation methods	Moisture levels	Total amount of water consumed (mm)
Drip	100% of PET	340
	125% of PET	425
	75% of PET	255
Sprinkler	100% of PET	340
	125% of PET	425
	75% of PET	255
Rain hose	100% of PET	340
	125% of PET	425
	75% of PET	255
BBF	100% of PET	340
	125% of PET	425
	75% of PET	255

Table 3.
Quantity of water applied through the different methods of irrigation at different moisture levels.

The quantity of irrigation provided to the different systems of irrigation would be the same, but the time taken by the system will vary according to their discharge rate. The amount of water provided through different methods of irrigation at different moisture stress levels is shown in **Table 3**.

The life irrigation was given to all the plots uniformly irrespective of irrigation scheduling treatment for the crop establishment, and the following irrigations were given as per the treatments through different irrigation methods.

The pod yield, oil yield, water indices, and economic indices were studied.

Water use efficiency:

WUE is the ratio between the yield obtained to the total amount of water consumed during the crop growth. The equation was given by [8] and expressed as kg/ha mm

$$WUE = \mathrm{Y} / \mathrm{W} \tag{2}$$

Where Y—yield (kg); W—total amount of water consumed (mm)

Water productivity:

Water productivity can be expressed in terms of returns earned from the seed (kernel) weight per amount of water consumed. It was expressed as Rs/ha mm

$$\text{Water productivity} = \frac{\text{Returns obtained}\left(Rs\right)}{\text{total amount of water consumed}\left(mm\right)} \tag{3}$$

The economics of the micro irrigation systems also had been derived to calculate the net returns and B:C ratio.

Cost of cultivation (Rs/ha):

The rupee invested from the operations of land preparation to harvesting the crop was calculated to evaluate the amount spent on raising the crop.

Gross return (Rs/ha):

The returns are earned by harvesting the crop. The local market price of the produce (pod yield) was computed to evaluate the gross return of the produce.

Net return (Rs/ha):

The net return was evaluated by deducting the cost of cultivation from the gross return.

$$Net\ \text{return} = \text{gross return} - \text{cost of cultivation} \qquad (4)$$

Benefit-cost ratio:

The B:C ratio value was calculated by dividing the gross return value by the cost of cultivation.

$$\text{B:C ratio} = \frac{\text{gross return}}{\text{cost of cultivation}} \qquad (5)$$

3.2.2 Results and discussion

The methods of irrigation at different moisture levels had a remarkable influence on the pod and oil yield. The individual and the combination treatment effects were briefly shown in **Table 4.**

Among all the irrigation methods, drip irrigation (I_1) attained the highest pod yield with 2849 kg/ha. It was preceded by rain hose irrigation (I_3) with 2760 kg/ha and further preceded by sprinkler irrigation I_2 with 2513 kg/ha, and the lowest pod yield was recorded in BBF (I_4) with 1978 kg/ha. Among all the moisture stress levels, 100% (S_1) of PET has a higher impact on pod yield with 2732 kg/ha, and it was closely followed by 75% (S_3) of PET with 2561 kg/ha and the lower level that was recorded at 125% (S_2) of PET 2282 kg/ha.

The interaction effect between the different methods of irrigation and different moisture levels registered a higher pod yield of 3167 kg/ha in drip irrigation at 100% (I_1S_1) of PET, and it was preceded by rain hose irrigation at 100% (I_3S_1) of PET (3037 kg/ha) and further preceded by drip irrigation at 75% (I_1S_3) of PET (2733 kg/ha).

Treatment	Pod yield (kg/ha)				Oil yield (kg/ha)			
	S_1	S_2	S_3	Mean	S_1	S_2	S_3	Mean
I_1	3167	2517	2733	2849	1137	817	876	943
I_2	2542	2450	2483	2513	861	783	786	810
I_3	3037	2479	2593	2760	1037	800	820	886
I_4	2143	1533	2103	1978	711	452	680	614
Mean	2732	2282	2561	2525	936	713	790	813
	I	S	I × S	S × I	I	S	I × S	S × I
S. Ed	77	70	141	146	7.63	8.66	17.33	20.36
CD (p = 0.05)	188	150	300	304	18.67	18.37	36.75	34.01

Table 4.
Effect of different methods of irrigation and different moisture levels on pod yield and oil yield of groundnut.

The combination of the BBF method of irrigation along with 125% (I_4S_2) of PET resulted in a lower pod yield of 1533 kg/ha.

Out of all the irrigation methods, drip irrigation I_1 expressed the highest oil yield (943 kg/ha). It was preceded by rain hose irrigation I_3 (886 kg/ha) and further followed by sprinkler irrigation I_2 (810 kg/ha), and the lower oil yield was registered by BBF I_4 of (614 kg/ha). Among all the moisture levels, 100% (S_1) of PET has a high oil yield (936 kg/ha), and it was closely followed by 75% (S_3) of PET (790 kg/ha), and the lower level of oil yield was present in 125% (S_2) of PET (713 kg/ha).

The interaction effect between the different methods of irrigation and different moisture levels had significantly influenced the oil yield. Drip irrigation at 100% (I_1S_1) of PET registered a high amount of oil yield (1137 kg/ha) and was followed by rain hose 100% (I_3S_1) of PET (1037 kg/ha) and further followed by drip irrigation at 75% (I_1S_3) (876 kg/ha). The combination of the BBF method of irrigation along with 125% (I_4S_2) of PET has low amount of oil yield (452 kg/ha).

3.2.2.1 Effect of treatments on the economic factors

The treatment effect on cost of cultivation, gross return, net return, and B:C ratio are presented in **Table 5**.

The rupee invested was higher under the sprinkler irrigation system at three different moisture stress levels of 100% (S_1), 125% (S_2), and 75% (S_3) of PET. The cost of cultivation was lesser in broad bed and furrow irrigation systems at three different moisture levels of 100% (S_1), 125% (S_2), and 75% (S_3) of PET.

The effect of different methods of irrigation at different moisture levels greatly influenced the gross return. The analysis of different methods of irrigation at different moisture levels revealed that drip irrigation at 100% (I_1S_1) of PET got the highest gross return of Rs. 190,000/ha among all other moisture stress levels and it was closely followed by rain hose irrigation at 100% (I_3S_1) of PET (Rs. 182,220/ha), and the next level of net return (Rs. 1,64,000/ha) was recorded in drip irrigation at 75%

Treatment	Cost of cultivation (Rs./ha)	Gross return (Rs./ha)	Net return (Rs./ha)	B:C ratio
I_1S_1	59,560	190,000	130,440	3.19
I_1S_2	59,560	151,000	91,440	2.54
I_1S_3	59,560	164,000	104,440	2.75
I_2S_1	56,353	152,520	96,167	2.71
I_2S_2	56,353	147,000	90,647	2.61
I_2S_3	56,353	149,000	92,647	2.64
I_3S_1	55,353	182,220	126,867	3.29
I_3S_2	55,353	148,760	93,407	2.69
I_3S_3	55,353	155,596	100,243	2.81
I_4S_1	52,603	128,600	75,997	2.44
I_4S_2	52,603	92,000	39,397	1.75
I_4S_3	52,603	126,200	73,597	2.40

Table 5.
Effect of different methods of irrigation and different moisture levels on economics of groundnut.

of PET. The lowest gross return of Rs. 92,000/ha was attained in the Broad Bed and Furrow (BBF) method of irrigation at moisture levels of 125% (I_4S_2) of PET.

The highest net return of (Rs. 130,440/ha) was attained by the treatment combinations of drip irrigation along with 100% (I_1S_1) of PET. It was closely followed by rain hose irrigation at 100% of PET (I_3S_1) (Rs. 126,867/ha) and further followed by drip irrigation at 75% (I_1S_3) of PET (Rs. 104,440/ha). The lowest net return of Rs. 39,397/ha was obtained in the broad bed irrigation (BBF) with 125% (I_4S_2) of PET.

The calculated B:C ratio was higher in the treatment of rain hose irrigation with the moisture level equal to the 100% (I_3S_1) PET was 3.29. It was followed by drip irrigation at 100% (I_1S_1) of PET (3.19), and further by rain hose irrigation at 75% (I_3S_3) of PET (2.81). The least B: C Ratio of (1.75) was obtained in the broad bed and furrow irrigation (BBF) with a moisture level of 125% (I_4S_2) of PET.

Drip irrigation at 100% of PET (I_1S_1) had recorded higher gross return and net return when compared to other irrigation treatments. The increased yield under drip irrigation at 100% of PET (I_1S_1) had directly influenced the treatment to attain high gross and return returns. Though the drip at 100% of PET (I_1S_1) attained higher gross and net return, whereas, the highest B:C ratio was obtained in rain hose irrigation at 100% of PET (I_3S_1) achieved the highest B:C ratio due to its low installment cost, and it was also the next best to drip irrigation at 100% of PET (I_1S_1) in all factors including pod yield, lower disease incidence, higher hundred pod weight, high oil yield, shelling percentage, and so forth. The drip irrigation at 100% of PET (I_1S_1) attained second next to the rain hose at 100% of PET (I_3S_1) in terms of B:C ratio, and it was followed by the rain hose irrigation at 75% of PET (I_3S_3). Thus, the rain hose acts as an alternate low-cost micro-irrigation system. Therefore, the lowest B:C ratio was recorded in BBF at 125% of PET (I_4S_2) due to its lower number of pods per plant, lower pod yield, and lower shelling percentage leading to lower pod yield.

3.2.2.2 Effect of treatments on water indices of groundnut

The influence of treatment on water use efficiency and water productivity is presented in **Table 6**.

Treatment	Water use efficiency (kg/ha mm)				Water productivity (Rs./ha mm)			
	S_1	S_2	S_3	Mean	S_1	S_2	S_3	Mean
I_1	9.31	5.92	12.15	9.13	558.82	355.29	728.89	547.67
I_2	7.48	5.76	11.04	8.09	470.29	345.88	662.22	484.07
I_3	8.93	5.83	11.53	8.76	535.94	350.02	691.54	525.83
I_4	6.3	3.61	9.35	6.42	378.24	216.47	560.89	385.20
Mean	8.01	5.28	11.02	8.10	479.28	316.92	660.89	485.69
	I	S	I × S	S x I	I	S	I × S	S × I
S. Ed	0.07	0.17	0.27	0.21	8.26	5.01	10.03	13.36
CD (p = 0.05)	0.17	0.29	0.58	0.43	20.20	10.63	21.26	24.36

Table 6.
Effect of treatments on pod water use efficiency and water productivity of groundnut.

3.2.2.2.1 Water use efficiency (kg/ha mm)

The different methods of irrigation at different moisture levels had highly influenced the water use efficiency on a yield basis. The individual and the combination treatment effects were briefly shown in **Table 6**.

Out of all the irrigation methods, drip irrigation I_1 had higher water use efficiency of 9.13 kg/ha mm. It was nearly followed by rain hose irrigation I_3 with 8.76 kg/ha mm and further followed by sprinkler irrigation I_2 with 8.09 kg/ha mm, and the lower water use efficiency was recorded in BBF I_4 with 6.42 kg/ha mm. Among all the moisture levels, 75% S_3 of PET achieved higher water use efficiency of 11.02 kg/ha mm. It was followed by 100% S_1 of PET with 8.01 kg/ha mm, and the lower level of water use efficiency (5.28 kg/ha mm) was recorded in 125% S_2 of PET.

The interaction between the different methods of irrigation and different moisture levels has a remarkable effect on water use efficiency. Drip irrigation at 75% I_1S_3 of PET (I_1S_3) had recorded higher water use efficiency of 12.15 kg/ha mm. It was followed by rain hose at 75% I_3S_3 of PET (11.53 kg/ha mm) and further followed by sprinkler irrigation at 75% I_2S_3 of PET (11.04 kg/ha mm). The combination of the BBF method of irrigation along with 125% I_4S_2 of PET (3.61 kg/ha) has not effectively utilized the water when compared to all other irrigation treatments and registered least water use efficiency of 3.61 kg/ha mm.

3.2.2.2.2 Water productivity (Rs./ha mm)

Out of all the irrigation methods, drip irrigation I_1 had higher water productivity of Rs. 547.67/ha mm. It was nearly followed by rain hose irrigation I_3 with Rs. 525.83/ha mm and further followed by sprinkler irrigation I_2 with 484.07/ha mm, and the lower water productivity was obtained in BBF I_4 with Rs. 385.20/ha mm. Among all the moisture levels, 75% S_3 of PET achieved higher water productivity of 660.89/ha mm. It was followed by 100% S_1 of PET with Rs. 479.28/ha mm, and the lower level of water productivity was recorded in 125% S_2 of PET (Rs. 316.92/ha mm).

The interaction between the different methods of irrigation and different moisture levels had a remarkable effect on water productivity. Drip irrigation at 75% I_1S_3 of PET (I_1S_3) has recorded higher water productivity of Rs. 728.89/ha mm. It was followed by rain hose at 75% I_3S_3 of PET (Rs. 691.54/ha mm) and further followed by sprinkler irrigation at 75% I_2S_3 of PET (Rs. 662.22/ha mm). The combination of the BBF method of irrigation along with 125% I_4S_2 of PET had registered least water productivity of Rs. 216.47/ha mm.

The reason might be higher yield provided per mm of water consumed through reduced plant and soil surface evaporation by providing optimum moisture to the crops by direct application of water to the root zone, thereby avoiding moisture stress and providing optimum moisture to the crops [9]. The lower water use efficiency was obtained in BBF at 125% of PET (I_4S_2). Here, a high quantity of water produced a lesser yield due to the utilization of a high amount of water for vegetative growth at the vegetative phase led to low pod yield. The lower utilization of water for the development of reproductive parts was due to higher consumption of water during the early vegetative phase leads to excessive vegetative growth so that yield might be reduced [10].

Similarly, the different methods of irrigation and different moisture levels significantly influenced the water use efficiency (**Figure 1**). The result of increased water

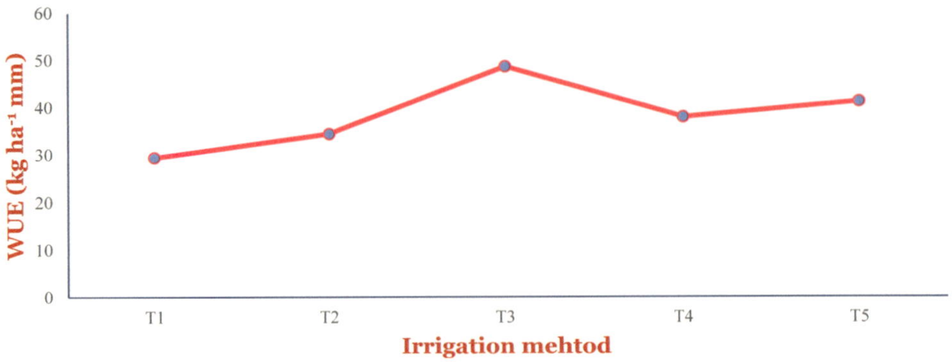

Figure 1.
WUE for different method of irrigation.

use efficiency had influenced the increased water productivity (**Figure 2**) by providing higher returns per amount of water consumed [11]. The lower water productivity was recorded in BBF at 125% of PET (I_4S_2). The probable reason could be lower water use efficiency resulted in lower water productivity by providing lower yield per quantity of water consumed [12].

The economic assessment also concluded that drip irrigation at 100% of PET (I_1S_1) has given the highest gross return, and it also attributed to the highest net return when compared to all other irrigation treatments. Though drip irrigation at 100% of PET (I_1S_1) (3.19) has obtained the highest gross and net return, the highest B:C ratio (3.29) was achieved in rain hose at 100% of PET (I_3S_1) because of its low material installation cost when compared to drip irrigation at 100% of PET (I_1S_1). The research concluded that drip irrigation at 100% of PET (I_1S_1) had performed well on all parameters and produced higher productivity. The rain hose at 100% of PET (I_3S_1) achieved higher profitability [13].

Rain hose irrigation system achieved higher yield of 1172 kg/ha with the BCR of 2.48 in black gram and 2760 kg/ha of pod yield with the BCR of 2.39 in groundnut.

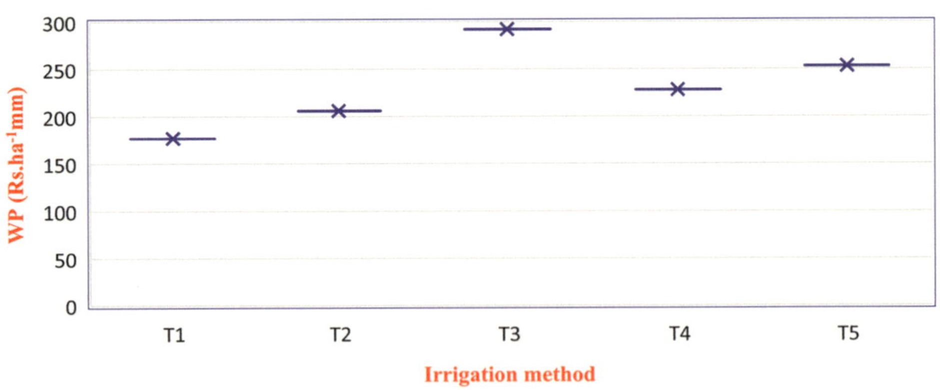

Figure 2.
WP for different method of irrigation.

4. Conclusion

The promotion of low-cost irrigation system supports for the achieving the yield potential of major agricultural closed spaced crops apart from enhancing the yield, water use efficiency, water productivity, net income, and BCR. The system components are easy to install, maintain, and operate with less knowledge and skill to achieve higher performance. Government supports to alternate low-cost irrigation system for certain crops reduce the water requirement and reduce the water demand of nation in irrigated agriculture.

Author details

Vallal Kannan Sankaralingam[1]*, Marimuthu Subramanian[2], Sangeetha Jebalin Vijayalalitha Vallal Kannan[3] and HariniHarishma Veerappan Prithviraj[3]

1 Programme Coordinator (ICAR-TNAU), Krishi Vigyan Kendra, Ramanathapuram, Tamil Nadu, India

2 National Pulses Research Centre (ICAR-TNAU), Pudhukkottai, Vamban, Tamil Nadu, India

3 Agricultural College and Research Institute, Tamil Nadu Agricultural University, Coimbatore, Tamil Nadu, India

*Address all correspondence to: vallalkannan@gmail.com

IntechOpen

© 2025 The Author(s). Licensee IntechOpen. This chapter is distributed under the terms of the Creative Commons Attribution License (http://creativecommons.org/licenses/by/4.0), which permits unrestricted use, distribution, and reproduction in any medium, provided the original work is properly cited.

References

[1] Indian Council of Agricultural Research (ICAR). Project Coordinators Report, All India Coordinated Research Project on Kharif Pulses. Kanpur: ICAR-Indian Institute of Pulses Research; 2023. pp. 1-114

[2] Mustafa Y, Grando S, Maatougui M. Optimal agronomic practices to improve water productivity for barley in highlands of Eritrea. In: Paper Presented at the CPWF Workshop on Increasing Water Productivity of Rainfed Cropping Systems. Tamale, Ghana. 2008

[3] Marimuthu S, Kannan SV, Pazhanivelan S, et al. Harnessing rain hose technology for water-saving sustainable irrigation and enhancing blackgram productivity in garden land. Scientific Reports. 2024;**14**:18692. DOI: 10.1038/s41598-024-69655-2

[4] Waseem M, Kaleel I, Mallikarjuna N, Polisgowsar B. Effect of micro sprinkler and surface irrigation on growth and yield of groundnut crop under Raichur agro-climatic conditions. Journal of Pharmacognosy and Phytochemistry. 2018;**7**(1):132-134

[5] Soni JK, Raja NA, Ajaykumar R. Response of physiological and yield attributing parameters of groundnut under drip and micro sprinkler fertigation. International Journal of Current Microbiology and Applied Sciences. 2017;**8**(6):2210-2216

[6] Swain O, Mohapatra P, Digal B, Sahu A. Water use efficiency of sunflower under deficit drip irrigation in east and south-east coastal plain agro-climatic zone of Odisha, India. International Journal of Current Microbiology and Applied Sciences. 2019;**8**(6):2210-2216

[7] Rathore VS, Nathawat NS, Bhardwaj S, Yadav BM, Kumar M, Santra P, et al. Optimization of deficit irrigation and nitrogen fertilizer management for peanut production in an arid region. Scientific Reports. 2021;**11**(1):1-14. DOI: 10.1038/s41598-021-84837-0

[8] Viets FG. Fertilizers and the efficient use of water. Advances in Agronomy. 1962;**14**:223-264

[9] El-Habbasha S, Okasha E, Abdelraouf R, Mohammed A. Effect of pressured irrigation systems, deficit irrigation and fertigation rates on yield, quality and water use efficiency of groundnut. International Journal of ChemTech Research. 2014;**15**(7):1

[10] Chai Q, Gan Y, Turner NC, Zhang RZ, Yang C, Niu Y, et al. Water-saving innovations in Chinese agriculture. Advances in Agronomy. 2014;**126**:149-201. DOI: 10.1016/B978-0-12-800137-0.00003-4

[11] Gadade G, Dhopte R, Khodke U. Effect of different spacing on growth and yield of BBF raised summer groundnut (*Arachis hypogaea* L.) under drip irrigation. International Journal of Current Microbiology and Applied Sciences. 2018;Special Issue;**6**:593-597

[12] Kannan N, Senthivel T. Integrated land and water management practices for hill side maize crop cultivation under rainfed condition at Rubirizi, Rwanda, East Africa. International Journal of Current Microbiology and Applied Sciences. Special Issue 6. 2018:593-597

[13] Harishma VH, Kannan SV, Sampathkumar T, Sheeba S. Yield enhancement through a low-cost irrigation system for groundnut. International Journal of Environment and Climate Change. 2022;**12**(11):1461-1466

Chapter 3

Perspective Chapter: Adopting Integrated Methods in Communicating Efficient Small-Scale Irrigation Practices and Dry-Season Farming in West Africa's Rural Communities

Agwu A. Ejem

Abstract

The persistent food insecurity in West Africa's rural communities, largely caused by the variability of rainfalls, highlights the urgent need for efficient small-scale irrigation practices and dry-season farming among smallholder farmers who constitute at least 80% of the farming population in the sub-region. This system has proven effective in enhancing agricultural productivity, ensuring a continuous supply of food all year round, improving the livelihoods of farmers and their communities, and overcoming the effect of climate change in the dry season. However, there are information and communication challenges that have impinge on the efficient adoption of these systems. This chapter suggested the use of integrated communication methods in driving the adoption of efficient small-scale irrigation practices and dry-season farming in West Africa's rural communities. It is hoped that this method will be suitable for these rural communities, which are home to diverse groups, including farmers, fishers, traders, and local leaders; it ensure that different audiences receive tailored information through channels they prefer, be it traditional media, digital platforms, or face-to-face interactions; and that information is delivered in various languages, ensuring inclusivity and better understanding, among others.

Keywords: dry-season farming, integrated communication methods, irrigation practices, small-scale irrigation; sustainable development goals, West Africa's rural communities

1. Introduction

The variability in rainfall in West Africa has adverse consequences on agricultural productivity and food security in the sub-region [1]. Despite regionally high rainfall,

IntechOpen

drought caused by climate change has accounted for the massive loss of farms and a concomitant extensive economic loss in many parts of the sub-region, threatening the achievement of the Sustainable Development Goals (particularly, goals 1, 2, 3, 12, and 13). Due to climate change, the impacts of droughts have been more severe in recent years [2]. As a result, food insecurity has remained a perennial problem in the regions. The problem persisted and has worsened, irrespective of previous and existing agricultural policies aimed at increasing food production and reducing hunger in West African countries [3, 4].

This situation has highlighted the need for efficient small-scale irrigation practices and dry-season farming, which have proven to be effective in enhancing agricultural productivity, building coping strategies, and overcoming the effect of climate change in the dry season or during floods. Small-scale irrigation also known as smallholder irrigation or farmer-led irrigation is self-supplied irrigation technology owned and managed by peasant farmers or small groups of farmers [5]. Effective utilization of low-cost small-scale irrigation improves food production as well as household income and nutrition [6, 7]. Dry-season farming in Nigeria entails that land preparation close to a water freshwater source is carried out between late September and October, followed by planting operations, which often start in late October and run until late March [8]. Success in dry-season agricultural production and productivity is facilitated by the use of drought-resistant crops as well as efficient and effective irrigation systems [5]. Experts have identified that a wide range of crops such as rice, maize, tomatoes, watermelon, garden egg, pumpkin, cucumber, carrots, peppers, sweet potatoes, onions, okra, spinach, eggplant, melon, and cotton can be planted in the dry season [9].

However, perennial food insecurity in West Africa has been largely caused by a lack of or poor adoption of irrigation practices and dry-season farming, especially among the smallholder farmers who constitute the largest population of farmers in the sub-region [3, 10]. Irrespective of the abundant irrigable water resources in West Africa, over-reliance on rainfed cultivation has led to a reduction in food production, seasonal food shortages, and hikes in food prices during the dry season. According to Ben-Enukora et al. [3], agricultural products become abundant during the rainy season, resulting in a surplus of food, crops, and vegetables and therefore causing prices to fall and farm produce to be wasted due to lack of storage facilities and the inability to participate in large-scale production in the dry-season results in food shortages and an increase in food prices.

The major factor responsible for the poor utilization of small-scale irrigation practices and dry-season farming is poor knowledge of the systems among smallholder farmers. Communication has been known to be essential in facilitating the awareness and adoption of small-scale irrigation practices and dry-season farming. Communication is an integral part of agriculture. It involves supplying appropriate knowledge and information flow to different actors in the agricultural sectors [11]. Thus, information is key to achieving an increase in the food supply, the best prices for food items throughout the year, reducing reliance on food importation, an increase in the county's GDP through food export, and improving the quality of land use [3]. Effective communication is among the significant factors in the efficient adoption of irrigation practices, dry-season farming, and flood risk management. Communicating the adoption of innovations in agriculture and climate change management is one of the main challenges that public management and other actors have in the search for the food security construction of resilient cities and societies, and regional development [12].

There is evidence that the adoption of small-scale irrigation practices and dry-season farming has been challenged by the communication method adopted in communicating them to the rural population who are largely non-literate, dispersed, and more vulnerable – being largely smallholder farmers. What is often deployed is top-down communication from official sources using the mainstream media. There is evidence that this method has many challenges and has rarely been successful. A survey by Ben-Enukora et al. [3] shows that the knowledge gained from the dry-season agricultural programs in the broadcast media is not sufficient for farmers to engage in dry-season irrigation farming successfully. Most notably, the communication method is inaccessible to many smallholder farmers and further widens the existing inequalities between those who can access the mainstream media and those who cannot. Moreover, it exacerbates the vulnerability of those who live in rural areas and cannot afford radio, TV, newspapers, social media, etc. Also, information from government agencies is sometimes received with skepticism, and as a result, people do not take necessary action toward small-scale irrigation practices and dry-season (or irrigated) farming.

This chapter will suggest the adoption of integrated methods that will resonate with the demographic and psychographic characteristics of the smallholder farmers in communicating efficient small-scale irrigation practices and dry-season farming in West Africa's rural coastal communities. This method will integrate traditional, modern, and digital communication systems, as well as adapt the messages to different cultural contexts and linguistic capabilities. It will also ensure that the audience is co-creators of messages on small-scale irrigation practices and dry-season farming, and that the mainstream sources of communication will be combined with indigenous methods to facilitate clarity and understanding.

2. Methods

This chapter's methodology is a semi-systematic review of the literature. The literature search was conducted on SCOPUS and Google Scholar. Both are free of charge and credible electronic databases. However, after the initial search, important webpages found in the references of the selected articles were searched on Google. The search made use of search terms, subject headings, and keywords, including "agricultural productivity," "dry-season farming," "food security," "integrated communication," "irrigation," "rural communities in West Africa," and "small-scale irrigation." Boolean operators ("AND" and "OR") and wildcards (* and /) were used to combine the search terms as a query in the databases. Inclusion criteria include articles published in English, published in 2017 and later, and full texts, empirical and observational studies (systematic review, opinion articles, mixed-method research, and literature review).

3. West Africa's rural communities

Understanding the unique characteristics and challenges of West Africa's rural communities, their need for efficient small-scale irrigation practices and dry-season farming, and the challenges that impinge on them from achieving high agricultural productivity can be better appreciated and effectively overcome. West Africa's rural communities have been explained as dynamic and resilient, with a deep connection to their cultural heritage and environment [13].

Geographically, West Africa's rural areas have varied landscapes that include forests, savannas, mountains, and coastal plains, and each of these areas offers unique challenges and opportunities for agriculture and daily living. Some rural areas, such as those in Southern Nigeria, are densely populated and have homesteads called compounds, while other areas have low population densities and live in settlements. Additionally, there is a consensus that the rural regions of West Africa span different climate zones, ranging from the arid Sahara in the north to the humid tropics in the south, and the climatic differences affect agricultural practices and living conditions in these areas [14].

Economically, the major occupations in West Africa's rural areas are farming, fishing, and trade/crafts. Farming is the backbone of the rural economies in the region. The most common crops cultivated include maize, cassava, millet, sorghum, and yams. Livestock rearing is also significant in many parts of the region, with the commonest being goats, cattle, and sheep [15]. Fishing is also significant in the region, especially in the coastal and riverine communities that rely on artisanal and commercial fishing for sustenance and income. As with trade and crafts, rural markets in West African rural communities bustle with trade in handmade crafts and other goods. The crafts that are traditional to most West African communities include pottery, weaving, and basket making.

Socio-culturally, West Africa's rural communities are home to many ethnic groups, each with its languages, dialects, customs, and traditions, and the cultural diversities are celebrated through rituals, festivals, and dance [16]. The community structure is built around social organizations, which often center around extended families, clans, and traditional leadership. The traditional authority is held by elders and community leaders who play crucial roles in decision-making.

In terms of infrastructure and services, many rural areas face road and transportation challenges, with inadequate road networks that limit access to farms, markets, communities, healthcare, and education. Access to clean water and proper sanitation have also been identified as some of the features of the West African rural communities, and this has led to health issues and has affected the overall well-being of many of these communities. Educational facilities in some of the rural communities may be under-resourced or scarce, and this has impacted the quality of education available to children and youth in many of these rural communities.

Environmentally, most of the dominant issues in these areas are climate change and land degradation. These rural communities are especially vulnerable to the impacts of climate change, such as floods, droughts, and erratic rainfall, which affect agricultural productivity and food security. In many parts of the region, deforestation, soil erosion, and overgrazing have led to land degradation, and this has reduced the quality and availability of arable land and has affected livelihoods in these communities.

How do West African rural communities build resilient and adaptation strategies? These communities often rely on traditional knowledge to manage their resources and cope with the challenges that beset their environment. This traditional knowledge is passed down through generations and is essential for the sustenance of these communities. Also, many of these communities are adopting (or are willing to adopt) innovative practices in the areas of sustainable agriculture, agroforestry, and community-based natural resources management to enhance their resilience and sustainability. The challenge is the poor spread of awareness and knowledge about these innovations and the lack of widespread adaptation in these communities.

DOI: http://dx.doi.org/10.5772/intechopen.1010668

4. Efficient irrigation practices

Irrigation is the process of artificially supplying water to agricultural fields to support crop growth. It has also been defined as the supply of controlled amounts of water to land at regular intervals through canals or other artificial methods to enhance agricultural growth in areas where rainfall is insufficient, irregular, or during dry seasons [17].

Different irrigation techniques are employed to ensure that crops receive the necessary water for optimal growth. The interest in this chapter is small-scale irrigation, which is reviewed below, along with other common irrigation techniques:

4.1 Small-scale irrigation

Smallholder irrigation is a type of irrigation that is used on small farms, often less than 2 hectares, that is privately owned and controlled by the farmer. Small-scale irrigation, also known as smallholder irrigation or farmer-led irrigation is a self-supplied irrigation technology owned and managed by peasant farmers or small groups of farmers [5]. Effective utilization of low-cost small-scale irrigation improves food production as well as household income and nutrition [6, 7]. Irrigation policy in many developing nations has shifted from large-scale public irrigation systems to private and small-scale irrigation due to the high cost of construction and maintenance of comprehensive irrigation projects [5].

Small-scale irrigation works by allowing farmers to choose which technologies to use to extract water, such as solar water pumps or manual lifting; select irrigation methods, such as by buckets or drip kits; and they can choose to purchase, run, and maintain the irrigation system themselves or with a small group of farmers. More so, small-scale irrigation can be adapted to the needs of each farmer and is more affordable and nimbler than larger irrigation systems, which can be delayed by construction.

4.2 Surface irrigation

Here, water is distributed over and across the land by gravity. This method includes techniques such as furrow, basin, and border irrigation. It is simple to implement and low-cost, but it can be inefficient due to water loss through evaporation and runoff and requires leveled fields.

4.3 Drip irrigation

Here, water is delivered directly to the root zone of plants through a network of valves, pipes, tubing, and emitters. It is highly efficient, minimizes water wastage, reduces weed growth, and can be automated. However, the initial setup cost is high, and emitters can clog if water is not filtered properly.

4.4 Sprinkler irrigation

Here, water is sprayed into the air and allowed to fall on crops like rain through a system of pipes and sprinklers. It allows for uniform water distribution, can cover a large area, and can be automated. However, high initial cost, wind can affect water distribution, and can cause leaf wetting, which may lead to diseases.

4.5 Subsurface irrigation

In this case, water is applied below the soil surface through buried pipes or drip tape. It reduces water evaporation and minimizes soil erosion; efficient use of water. However, high installation costs and maintenance can be challenging.

4.6 Manual irrigation

Here, water is manually carried and applied to crops using buckets, watering cans, or other containers. Its advantage is that it is low-cost and simple to use. However, it is Labor-intensive and not suitable for large-scale farming.

4.7 Center pivot irrigation

It uses a rotating sprinkler system that pivots around a central point, watering crops in a circular pattern. It is efficient for large fields, can be automated, and provides uniform water distribution. Its advantage is its high initial investment and maintenance cost, and it is not suitable for irregularly shaped fields.

4.8 Flood irrigation

Here, water is released onto fields to flood them, allowing it to soak into the soil. It is simple and low-cost, but inefficient water use can lead to waterlogging and soil erosion.

4.9 Lateral move irrigation

This is similar to a center pivot but moves laterally across the field. It is suitable for rectangular fields and uniform water application. But it requires a flat field and a high initial cost.

5. Importance of efficient irrigation system in West Africa's rural communities

Efficient irrigation is critical for sustainable agriculture in West Africa's rural communities because of some of the reasons that have been explained earlier, especially with scarce and erratic water resources.

Here are some of the importance of efficient irrigation systems:

1. It ensures that water is used precisely where and where it is needed and reduces wastage from runoff and evaporation.

2. Farmers can also maintain the long-term availability of water, especially in areas that are prone to drought.

3. It increases yields by providing crops with the right amount of water and the right time to promote crop growth and maximize potential yield.

4. It also enhances the quality of produce and leads to better marketability and higher prices.

5. It reduces the need for fertilizers and pesticides and lowers the overall cost of production.

6. It makes farming more economically viable by reducing the cost of water supply.

7. It preserves soil health and structure and prevents the excessive flow of water that can wash away topsoil.

8. It helps farmers to adapt to changes in weather patterns by ensuring a reliable water supply during dry spells.

9. It reduces the carbon footprints of the agricultural sector and contributes to climate resilience.

10. It supports year-round farming, including during the dry season, and ensures a steady supply of food.

11. It reduces the risk of crop failures, leading to food security.

12. Framers are empowered to improve their livelihoods and economic status.

6. Dry-season farming

Dry-season farming refers to agricultural practices carried out during the dry season when natural rainfall is minimal or nonexistent. This method is particularly important in regions with distinct wet and dry seasons, like West Africa, where maximizing the growing season is essential for food security and economic stability. It works hand-in-hand with irrigation, which is essential to providing water to crops at that time. Other than irrigation, other techniques are that mulching is done to retain moisture, regulate soil temperature, and suppress weeds at that time, and certain drought-resistant or fast-maturing crops are plated at that time. According to Ben-Enukora et al., a wide range of crops such as rice, maize, tomatoes, watermelon, garden egg, pumpkin, cucumber, carrots, peppers, sweet potatoes, onions, okra, spinach, eggplant, melon, and cotton can be planted in the dry season [9].

Dry-season farming entails that land preparation close to a water freshwater source is carried out between late September and October, followed by planting operations, which often start in late October and run until late March [6]. Among other benefits, which will be reviewed below, dry-season farming is profitable as it yields high agricultural output [3, 5]. However, success in dry-season agricultural production and productivity is facilitated by the use of drought-resistant crops as well as efficient and effective irrigation systems [5].

Importance of dry-season farming

1. It ensures a continuous supply of food all year round.

2. It provides farmers with an opportunity to generate income during the dry season.

3. It helps farmers to maximally utilize land and resources.

4. It encourages the adoption of sustainable farming practices that improve soil health and conserve water.

5. It helps farmers to adapt to changing climate conditions by ensuring the production of foods in the absence of rain.

6. It can help reduce rural-urban migration and keep rural communities busy and intact.

7. It helps to stabilize prices and supports local economies by ensuring a consistent supply of agricultural products in the market.

7. Communication challenges facing efficient small-scale irrigation practices and dry-season farming in West Africa's rural communities

There are many challenges facing dry-season farming and small-scale irrigation in West Africa such as water availability and lack of proper irrigation infrastructure, among others. However, many studies have highlighted that the major factor responsible for the poor utilization of small-scale irrigation practices and dry-season farming is poor knowledge of these practices and a lack of awareness of how they can be used to achieve optimal results among smallholder farmers [3, 18].

Communication is an integral part of agriculture. It involves supplying appropriate knowledge and information to different actors in the sector [11]. Effective communication is among the significant factors in the efficient adoption of agricultural practices. It is, therefore, safe to say that communication is essential in facilitating the awareness and adoption of small-scale irrigation practices and dry-season farming. Poor communication of innovations in agriculture and climate change management is one of the main challenges that the government and other actors have in the search for food security and in constructing resilient cities and societies [12].

The communication challenges facing efficient small-scale irrigation practices and dry-season farming in West Africa's rural communities include the following:

a. Limited access to information. The broadcast media have been useful in disseminating agricultural policies and practices such as dry-season farming and irrigation, and they are known to reach a wide audience, but many farmers in rural West African areas may not have regular access to television or radio programs. More so, the lack of internet and digital literacy in the West African rural communities may also hinder the spread of information on agricultural policies and practices such as dry-season farming and irrigation [19].

b. Ineffective communication strategies. The media information on these agricultural practices is sometimes not available in local languages, and farmers find it difficult to understand and apply the knowledge. Research has also shown that broadcast programs on agricultural practices are often aired when farmers are not listening or watching, thereby reducing their impacts [3, 19].

c. The extension service is challenged by a number of factors. There are a lot of challenges facing the agricultural extension services in parts of West Africa, prominent among them are shortage of agents to provide hands-on training and support the teeming farmers in agricultural practices, gross inadequate and untimely funding, to-down, poor conditions of service, supply drive approaches, poor targeting of vulnerable groups, and negative political influences, among many others [20]. The consequence of this is that farmers may not receive adequate training on best agricultural practices.

d. Poor engagement of stakeholders. Lack of engagement with local communities can result in a disconnect between the information provided and the actual needs of the farmers [19].

e. Regular use of formal channels and official sources. Many farmers will prefer communication channels that they can readily identify with such as the West African traditional communication system, and they want these messages to be delivered in language and manner that they are more familiar with [11, 21].

f. Financial challenges. The cost of accessing information through various channels can be prohibitive for smallholder farmers. More so, there is insufficient investment in communication infrastructure to ensure that information reaches all farmers effectively.

By overcoming these barriers, farmers can better adopt dry-season farming practices and improve their agricultural productivity.

8. Framework for integrated communication strategies in driving the adoption of efficient small-scale irrigation practices and dry-season farming in West Africa's rural communities

Addressing the challenges outlined above requires a multi-faceted approach that improves access to information and is created with the audience in mind. In this section, this chapter proposes a new framework for integrated communication strategies to drive the adoption of efficient small-scale irrigation practices and dry-season farming in West Africa's rural communities. The multi-faceted framework ensures that there is a channel for every category of audience – the one that addresses their language, reach, literacy level, time of exposure, preferred means of delivery, etc. It is called an integrated communication method.

This framework in **Figure 1** is a strategic approach that combines various communication methods and channels, such as digital, print, broadcast, face-to-face, and social media, to deliver a consistent message across all platforms. The integrated communication methods use traditional, modern, and digital communication systems, and adapt them to different cultural contexts and linguistic capabilities.

The integrated communication method in the framework below uses the same message but is delivered in the following ways (**Table 1**).

The integrated framework above was designed by the author in consideration of the evidence in various bodies of literature that West Africa's rural communities are home to diverse groups, including crop and livestock farmers, fishers, traders, blacksmiths, wood collectors, and local leaders [22, 23] and diverse demographics, such as literates, non-literates, elderly, youths, rich, middle-class, and poor [24, 25]. Therefore, the author considered a model that ensures that these different audiences receive tailored information through channels they prefer, be it traditional media,

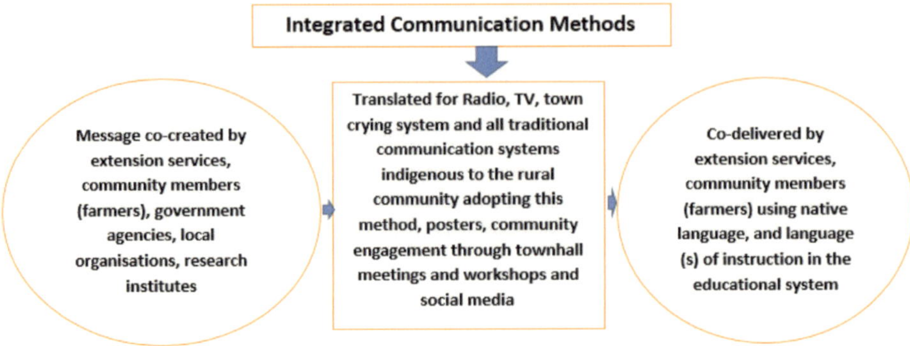

Figure 1.
Integrated communication framework for small-scale irrigation practices and dry-season farming in West Africa's rural communities (Author).

Channels/sources	Description
Conventional mass media	The method will use radio, television, and print media to reach a broad audience who can afford them, and the programs can include interviews and educational segments.
Digital media	The method will use social media platforms, mobile apps, and websites for those who are literate, to share information, resources, and updates. SMS and voice messaging can also be used. This enables two-way communication.
Indigenous/ traditional communication systems	West African communities have an indigenous system of communicating information to the people. This method will include the town-crying system, the age-grade system, folk media, festivals, gongs, and open markets, among others.
Interpersonal channels	*Community-based engagements:* These involve community meetings, workshops, and training sessions to share knowledge, demonstrate techniques, and address concerns. These engagements involve local leaders and influencers to encourage participation. They can also be farmer field schools where farmers can learn and practice new farming techniques, and peer-to-peer learning and knowledge sharing. *Extension services:* Use extension agents to provide personalized support and guidance to farmers.
Source of the messages	Use communities as co-creators of the message: Involve farmers and other stakeholders in the communities in the creation, planning, and implementation of communication strategies. This ensures that the information is relevant and tailored to the community's needs.

Table 1.
Communication channels/sources and modes of message creation and delivery (Author).

digital platforms, or face-to-face interactions. In addition, evidence in the literature shows that these communities often speak multiple languages and dialects [16]. Therefore, the author incorporated integrated strategies that can deliver information in various languages, ensuring inclusivity and better understanding.

Research has also shown that many of West Africa's rural communities are in remote locations with limited access to certain types of media [26]. By using a mix of communication channels, including radio, mobile apps, and village meetings, information can be disseminated effectively to even the most isolated areas. More so, while some community members may have access to digital platforms, others might rely on traditional methods; integrated communication bridges this gap by combining both digital and traditional approaches.

9. Importance of using integrated communication strategies in driving the adoption of efficient small-scale irrigation practices and dry-season farming in West Africa's rural communities

Integrated communication strategies are essential for the effective dissemination of information, engagement of stakeholders, and promotion of understanding among the audience, irrespective of their characteristics, preferences, and time of exposure. Using multiple communication channels ensures that information reaches a broad audience, including those in remote and rural areas. Also, it can provide farmers and communities with timely updates on weather, pest alerts, and best practices.

Integrated communication strategies are not new in agricultural systems in Europe, and have proven to be very effective in sustainable agricultural solutions in the continent. The European Commission has carried out several initiatives to address knowledge and innovation systems in the agricultural sector, and one of those initiatives is the adoption of traditional and new methods to engage with farmers. There are also activities organized by farmers, farmers' organizations, research institutes, extension services, NGOs, and other coalitions to facilitate learning [27]. Engaging the community in the planning and implementation of communication strategies ensures that their needs and concerns are addressed. Research has shown the importance of people-centered strategies in responding to agricultural and environmental challenges [28]. Additionally, workshops, field schools, and training sessions provide farmers with practical knowledge and skills to adopt new techniques and practices.

Integrated communication strategies have been shown to facilitate the sharing of knowledge and experiences among farmers, leading to the adoption of best practices and innovations in many countries. Also, it provides easy access to information on efficient irrigation and dry-season farming. Furthermore, everyone is carried along because every farmer will be exposed to the messages in one way or another. For instance, evidence has shown that India integrates modern media such as TV, radio, mobile phones, YouTube videos, PowerPoint presentations, and informative DVDs to impart information and skills training among farmers. They also use interpersonal communication, visits, and training to increase the intensity of the training [29]. It has also been proven in Nepal that a combination of mass media and interpersonal and traditional methods of communication such as group discussions, farmer field schools, folk media, visual aids, and group discussions not only raise awareness of sustainable agricultural practices among farmers but also ensures wide coverage, provide opportunities for in-depth learning and presents information in a culturally relevant and accessible formats [30].

Integrated communication strategies allow for the monitoring of progress and the evaluation of the effectiveness of various initiatives, and feedback from this monitoring exercise will provide valuable insights that can be used to refine and improve communication efforts.

10. Conclusion and recommendations

The persistent food insecurity in West Africa has been largely caused by a lack of or poor adoption of irrigation practices and dry-season farming, especially among the smallholder farmers who constitute the largest population of farmers in the sub-region. The major factor responsible for the poor utilization of small-scale irrigation practices and dry-season farming is poor knowledge of the systems among smallholder farmers. The onus is now on adequately communicating the adoption of these systems in West Africa's rural communities. However, the communication challenges facing efficient small-scale irrigation practices and dry-season farming in West Africa's rural communities include limited access to information, ineffective communication strategies, lack of extension services, poor engagement of stakeholders, and regular use of formal channels and official sources, among others. An integrated communication method was suggested for delivering the messages through the following ways: multi-channel communication, community-based engagements, use of extension agents to provide personalized support and guidance to farmers, and use of the Indigenous communication systems, among others.

It is therefore recommended here that to implement integrated communication methods in West African rural communities, the following steps should be taken:

For policymakers: conduct a thorough assessment of the communication needs, challenges, and preferences of the target populations; identify existing communication channels and their challenges; build partnerships with community leaders and other key stakeholders; choose a mix of communication channels; develop relevant, clear, and culturally appropriate messages; and ensure that the audience can access the information in local languages and dialects.

For extension workers and researchers: This model should be test-run in a rural community in West Africa, and the medium-term and long-term effectiveness in enhancing the knowledge of smallholder farmers in irrigation practices and dry-season farming should be evaluated. Also, future research directions should look at the alliance between farmers, extension workers, NGOs, research institutes, and policymakers in co-creating and delivering knowledge and innovation systems in the agricultural sector.

Author details

Agwu A. Ejem
Department of Mass Communication, Landmark University, Omu Aran, Nigeria

*Address all correspondence to: ejem.agwu@lmu.edu.ng

IntechOpen

© 2025 The Author(s). Licensee IntechOpen. This chapter is distributed under the terms of the Creative Commons Attribution License (http://creativecommons.org/licenses/by/4.0), which permits unrestricted use, distribution, and reproduction in any medium, provided the original work is properly cited.

References

[1] Adekunle OA, Oladipo FO, Busari IZ. Factors affecting farmers' participation in irrigation schemes of the lower Niger River basin and rural development authority, Kwara state, Nigeria. South African Journal of Agricultural Extension. 2015;**43**(2):42-51

[2] Vicente-Serrano SM, Peña-Angulo D, Beguería S, Domínguez-Castro F, Tomás-Burguera M, Noguera I, et al. Global drought trends and future projections. Philosophical Transactions of the Royal Society A. 2022;**380**(2238):20210285

[3] Ben-Enukora CA, Ejem AA, Aremu CO, Adeyeye BK, Oloruntoba AF. Access to dry season agricultural content in the broadcast media and dry season irrigation farming among smallholder farmers in Nigeria. Agriculture. 2023;**13**(6):1139

[4] Bjornlund V, Bjornlund H, van Rooyen A. Why food insecurity persists in sub-Saharan Africa: A review of existing evidence. Food Security. 2022;**14**(4):845-864

[5] Xie H, You L, Dile YT, Worqlul AW, Bizimana JC, Srinivasan R, et al. Mapping development potential of dry-season small-scale irrigation in sub-Saharan African countries under joint biophysical and economic constraints-an agent-based modeling approach with an application to Ethiopia. Agricultural Systems. 2021;**186**:102987

[6] African Union. Framework for Irrigation Development and Agricultural Water Management in Africa. 2020. Available from: https://au.int/en/documents/20200601/framework-irrigation-development-and-agricultural-water-management-africa [Accessed: December 3, 2024]

[7] Tefera E, Cho YB. Contribution of small-scale irrigation to households income and food security: Evidence from Ketar irrigation scheme, Arsi zone, Oromiya region, Ethiopia. African Journal of Business Management. 2017;**11**(3):57-68

[8] Farmers Square. Dry season farming in Nigeria. 2021. Available from: https://farmsquare.ng/drysea-son-farming-in-nigeria/#:~:text=Nigeria's%20dry%20season%20agriculture%20starts,agriculture%20to%20increase%20their%20income [Accessed: December 6, 2024]

[9] Ates S, Cicek H, Bell LW, Norman HC, Mayberry DE, Kassam S, et al. Sustainable development of smallholder crop-livestock farming in developing countries. IOP Conference Series: Earth and Environmental Science. 2018;**142**:012076

[10] Echendu AJ. Flooding, food security and the sustainable development goals in Nigeria: An assemblage and systems thinking approach. Social Sciences. 2022;**11**(2):59

[11] Marcillo-Delgado JC, Alvarez-Garcia A, García-Carrillo A. Communication strategies on risk and disaster management in South American countries. International Journal of Disaster Risk Reduction. 2022;**76**:102982

[12] Fonseca MN, Silva LP, Garcias CM. Flood risk communication. Mercator (Fortaleza). 2023;**21**:e21027

[13] Ouédraogo K, Zaré A, Korbéogo G, Ouédraogo O, Linstädter A. Resilience strategies of West African pastoralists in response to scarce forage resources. Pastoralism. 2021;**11**(1):16

[14] Adaawen S. Understanding climate change and drought perceptions, impact and responses in the rural Savannah, West Africa. Atmosphere. 2021;**12**(5):594

[15] Cocker F, Alamou EA, Elegbede BM, Toko II. Degree of integrated water resources management implementation in context of climate change in a watershed: Case of Oueme Basin, Benin (West Africa). Hydro Science & Marine Engineering. 2021;**3**(2):10-17

[16] Falola T. A History of West Africa. Cambridge: Taylor & Francis; 2023. DOI: 10.1109/SEB4SDG60871.2024.10630402

[17] Wang T, Wang Z, Guo L, Zhang J, Li W, He H, et al. Experiences and challenges of agricultural development in an artificial oasis_ A review. Agricultural Systems. 2021;**193**:103220

[18] Johnson JM, Becker M, Dossou-Yovo ER, Saito K. Farmers' perception and management of water scarcity in irrigated rice-based systems in dry climatic zones of West Africa. Agronomy for Sustainable Development. 2023;**43**(2):32

[19] Ejem AA, Aremu C, Ajakaiye OO, Ben-Enukora C, Akerele-Popoola OE, Ibiwoye TI, et al. Perspectives on communicating 21st-century agricultural innovations to Nigerian rural farmers. Journal of Agriculture and Food Research. 2023;**11**:100511

[20] Agwu AE, Suvedi M, Chanza C, Davis K, Oywaya-Nkurumwa A, Najjingo Mangheni M, et al. Agricultural extension and advisory Services in Nigeria, Malawi, South Africa, Uganda, and Kenya. In: Partnerships for Innovative Research in Africa (PIRA) Research Report. East Lansing, Michigan, USA: Alliance for African Partnership, Michigan State University; 2023

[21] Ben-Enukora CA, Soroaye MP, Ejem AA, Asogwa CE, Ojih SE, Ezegwu DT. Correlation between post-pandemic climate change advocacy on TikTok and knowledge, attitude, and adaptation practices among TikTok users in Nigeria. Journalism and Media. 2025;**6**(1):36

[22] Ahn J, Briers G, Baker M, Price E, Sohoulande Djebou DC, Strong R, et al. Food security and agricultural challenges in west-African rural communities: A machine learning analysis. International Journal of Food Properties. 2022;**25**(1):827-844

[23] Tarchiani V, Rossi F, Camacho J, Stefanski R, Mian KA, Pokperlaar DS, et al. Smallholder farmers facing climate change in West Africa: Decision-making between innovation and tradition. Journal of Innovation Economics & Management. 2017;**24**(3):151-176

[24] Pereira L, Mutesa L, Tindana P, Ramsay M. African genetic diversity and adaptation inform a precision medicine agenda. Nature Reviews Genetics. 2021;**22**(5):284-306

[25] Ineji BU. Trado-modern communication in development programmes: The social and behaviour change mix-media approach to mobilisation against breast cancer in rural Nigeria. CRUTECH Journal of Communication. 2024;**5**(1):55-58

[26] Sorgho R, Quiñonez CA, Louis VR, Winkler V, Dambach P, Sauerborn R, et al. Climate change policies in 16 West African countries: A systematic review of adaptation with a focus on agriculture, food security, and nutrition. International Journal of Environmental Research and Public Health. 2020;**17**(23):8897

[27] Sutherland LA, Marchand F. On-farm demonstration: Enabling

peer-to-peer learning. The Journal of Agricultural Education and Extension. 2021;**27**(5):573-590

[28] Ben-Enukora CA, Ejem AA. Environmental responsiveness towards desertification and land degradation: A review of literature on restoration strategies in dryland communities in Africa. In: 2024 International Conference on Science, Engineering and Business for Driving Sustainable Development Goals (SEB4SDG). IEEE; 2024. pp. 1-9

[29] Kaur A. Agricultural communication: A theoretical perspective. International Journal of Advanced Mass Communication and Journalism. 2022;**3**(1):73-77

[30] Dhakal A. The need for a multi-media approach in agricultural communication: Bridging socioeconomic gaps. European Journal of Theoretical and Applied Sciences. 2025;**3**(1):3-10

Chapter 4

Soilless Techniques and Innovative Solutions for Water-Efficient and Sustainable Agriculture

Tarek Mahrous and Ali Mohammed Abd-Elkader

Abstract

As global populations continue to rise, challenges such as increased food demand, climate change, water scarcity, insufficient fertile land, the spread of arid zones, and desertification are raised. These challenges have exerted unparalleled strain on conventional agricultural systems. Soilless culture techniques, including hydroponics, aeroponics, and aquaponics, have emerged as innovative and sustainable solutions to address these challenges. Soilless culture techniques transform modern agriculture by cultivating crops using nutrient-rich ions instead of soil. These methods offer a promising alternative to traditional agriculture by enabling efficient crop production in urban and non-arable areas, reducing reliance on synthetic fertilizers, and utilizing unconventional nutrient sources, thereby contributing to a biocircular economy. This chapter focuses on the potential of soilless cultivation to promote organic production, minimize pesticide use, and enhance resource efficiency, aligning with the Sustainable Development Goals (SDGs). Furthermore, it emphasizes the integration of advanced technologies such as artificial intelligence (AI) and the Internet of Things (IoT) for precise monitoring, decision-making, and optimization of resource use. The chapter also examines the economic viability, energy consumption, and cost-benefit analysis of these systems, emphasizing their role in transforming urban zones into eco-friendly and food-secure environments. By drawing attention to the multifaceted benefits of soilless culture techniques, this chapter aims to inspire broader adoption and investment in these technologies as a pathway toward a more sustainable and resilient agricultural future.

Keywords: soilless cultivation, hydroponics, aeroponics, water conservation, sustainable agriculture, cutting-edge technology, climate change, and environmental control

1. Introduction

Human activities have exacerbated climate change and water scarcity, posing significant threats to global food security as the world population continues to grow. Besides, water scarcity is a major challenge affecting over 2.4 billion people in water-stressed countries [1, 2].

Ensuring food security in the face of rising populations, climate change, and land degradation are among the most pressing global challenges [3]. Furthermore, there is another problem, which is arid zones that cover >40% of the global land and shelter over 40% of the world's population, besides the desertification and land degradation further aggravating environmental quality and food production in these areas [4]. Efforts to address these challenges have led to the development of sustainable food systems for drylands, focusing on optimizing water and nutrient efficiency. Soilless production methods, such as hydroponics, aeroponics, and aquaponics, offer a promising solution in areas where water and fertile land are limited [5]. Moreover, decreasing the dependency of soilless cultivation systems on synthetic fertilizers would enhance the sustainability of these systems [6].

Furthermore, one viable approach is to leverage agricultural waste and nutrient-rich effluents as substitutes for chemical fertilizers [7]. Likewise, real-time monitoring and data acquisition systems powered by AI are promising approaches in soilless culture techniques, enabling precise control over environmental parameters such as temperature, pH, and nutrient levels, thereby optimizing plant growth and resource efficiency in hydroponic and aeroponic systems. This approach uses IoT sensors and AI algorithms to analyze data and provide actionable insights, allowing for automated adjustments to ensure optimal growing conditions and reduce waste [8–10].

Soilless cultivation systems such as hydroponics, which cultivate plants in controlled soilless environments, allow for year-round production, particularly in space-constrained areas like urban or arid regions [5]. Studies have shown that hydroponics can accelerate plant growth by 30–500% compared to conventional farming methods, leading to higher yields and decreased fertilizer usage [11, 12]. Additionally, soilless culture techniques can decrease land and water requirements by up to 75 and 90%, respectively, offering a possible solution to global food scarcity [13]. Following a period of stagnation, hydroponics systems experienced significant growth between 2017 and 2024, with their agricultural application increasing by about 60% and their production value reaching approximately USD 5.4 billion, as shown in **Figure A1**. This trend is projected to double by 2030, driven by countries such as the United States, China, Canada, and Brazil. Furthermore, experts expect the worldwide hydroponics industry to grow at an annual rate of 11.9% from 2024 to 2030 [14].

Hence, the specific objectives of this chapter were to define and draw attention to soilless culture techniques as a sustainable solution for major global problems like food security, water shortage, and climate change. Furthermore, it introduces the different types of soilless techniques, how they efficiently use water, costs and benefits, and energy consumption. Besides the mentioned, the advancing monitoring technology in these systems, such as IoT and artificial intelligence (AI), is used to achieve optimization and full control for high-quality production. The above-mentioned suitability considerations and eco-friendly trends exploit waste (wastewater, digestate, etc.) as a nutrient source that supports a biocircular economy and achieves sustainability goals.

Our chapter literature review targeted peer-reviewed studies (2000–2024) from Scopus, Web of Science, and FAO technical reports, with particular attention to water scarcity [1, 2]. Search terms combined soilless techniques (hydroponics, aeroponics, and aquaponics), sustainability metrics (water efficiency and circular economy), and technologies (sensors, AI, data acquisition, and IoT), building on frameworks by Resh [5] and Goddek et al. [11]. We excluded non-empirical studies and prioritized data on unconventional nutrient sources (e.g., digestate [15] and wastewater [15]), aligning with the UN's food-water-energy nexus [13, 16]. From 1500 records, 120 studies met inclusion criteria after PRISMA screening (**Figure A2**), with emphasis on controlled environment agriculture [17] and urban applications [18].

2. Advanced soilless systems

Soilless culture is known as 'the technique of growing plants without the use of soil as a rooting medium', in which the nutrients absorbed by the plant roots are provided *via* water or nutrient solutions [19]. It is regarded as advanced practice; however, growing crops with comparable techniques have been tried through eras, probably primarily by Egyptians hundreds of years BC [20]. Soilless systems offer a significant alternative to soil farming in case of soil and/or water issues, such as water shortage and salinization [21]. Soilless culture techniques offer diverse methods of cultivating plants, each with its unique advantages and applications. In this classification, we explore some of the prominent techniques, shedding light on their distinct features and contributions to modern agriculture.

2.1 Hydroponics

Hydroponics, often known as hydroponics, is a modern technique for growing plants and crop production in a solution of water and mineral nutrition with no soil utilization [22]. The term 'hydroponics' refers to the Greek word 'hydro', which means water, and 'ponic', which means labor [23]. Generally, there are two techniques of hydroponics based on the media or substrate used: solution culture technique, where the plant roots are directly contacted with liquid nutrients, and granular-substrate or media culture technique, where the plants grow in an inert solid medium that provides support for the roots [24]. The solid media could be peat moss, charcoal, gravel, rock wool, perlite, coco peat, or coconut coir [25]. The hydroponics solution technique can be further categorized into two different types: continuous flow solution culture, where a pump is used to circulate the nutrient solution in the root zone and the surplus solution is collected and reused, such as nutrient film technique (NFT) and deep flow technique (DFT), as shown in **Figure 1**. And static solution culture refers to a method where nutrient solutions are not circulated and are provided only once when the electrical conductivity (EC) varies. These systems are named the root dipping method, the floating method, and wick watering [23].

Furthermore, hydroponic systems can be classified as open or closed systems depending on the control of the nutrient solution, as shown in **Figure 2**. In open systems, the nutrient solution, when applied to the plant roots, any excess discarded, while in closed systems recirculate and reuse the nutrient solution. The open system is simpler to manage; however, it is less efficient in water and nutrient use. On the other hand, closed systems offer significant water and nutrient savings but require more careful monitoring and management of the nutrient solution [26]. Additionally, closed systems have been found to achieve 90% water savings compared to open systems, with only 85%, besides saving fertilizers by 85% compared to open systems, which is only 68%. The above increase in crop productivity in the closed system was 300% compared to the open system (only 200%), and water productivity was 3500% in the closed system compared to 2000% in the open system [23]. However, closed systems may be more prone to the accumulation of harmful substances and imbalances in nutrient concentrations over time, necessitating regular monitoring and adjustment of the nutrient solution [27].

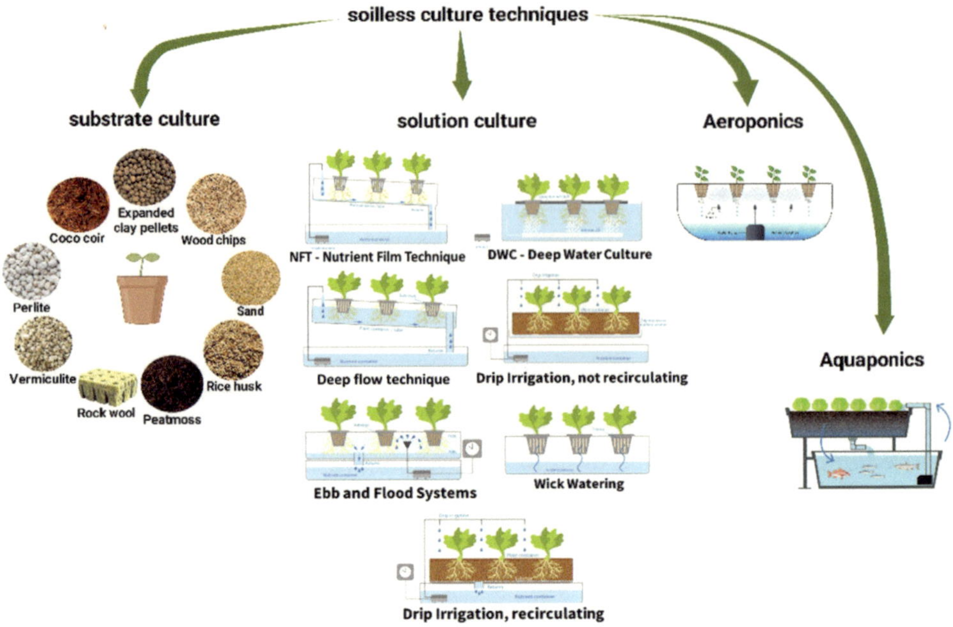

Figure 1.
Classification and schematic representation of major soilless cultivation techniques. The diagram illustrates three primary systems: (1) substrate culture, (2) solution culture, and (3) aeroponics/aquaponics. Key components are labeled to show differences in root zone environments, water/nutrient delivery mechanisms, and system architecture. This visual supports the technical discussion in Sections 2.1–2.3.

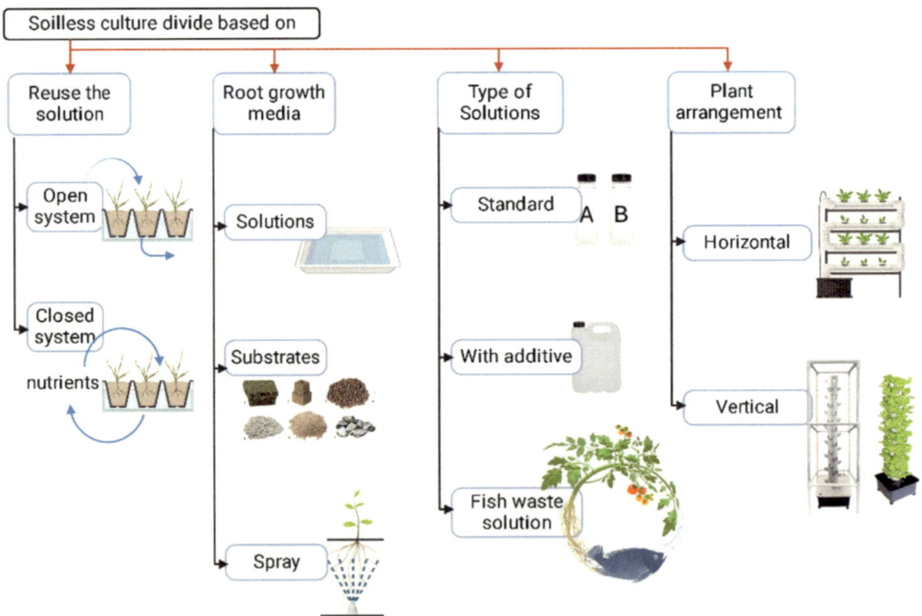

Figure 2.
Classification of soilless culture techniques according to reuse of nutrient solution, growth media, type of solution, and plant arrangement. The figure illustrates four primary systems: (1) according to reuse of solution recycle (closed) and non-recycle (open), (2) root growth media, (3) type of solution, and (4) plant arrangement.

DOI: http://dx.doi.org/10.5772/intechopen.1010669

Furthermore, **Table 1** introduces a comparative analysis of soilless cultivation systems: definition, advantages, limitations, and key references. Systems evaluated include nutrient film technique (NFT), deep water culture (DWC), Ebb and flood system (EFS), wick watering system (WWS), drip hydroponics system (DHS), aquaponics system, and aeroponics, with metrics on water savings and yield impacts.

Commercial and residential hydroponic systems are used to grow a wide range of plants, such as fruits, vegetables, flowers, and herbs, as shown in **Table 2**. Although the idea of soilless gardening has existed for thousands of years, the contemporary hydroponic system was created in the mid-1900s as a way to increase food production in areas with little space and resources [18].

Soilless system	Definition	Merits	Demerits	References
Nutrient film technique (NFT)	Suspended plant roots in a slightly sloping growing channel have been supplied with a continuous nutrient flow solution	• Less water consumption by 70–90% compared to the soil • No clogging • Abundant oxygen flow	• Weak supply of nutrients to the plant roots due to low flow rate • Any disruption in pumps leads to crop deterioration and yield loss	[28]
Deep water culture (DWC)	This system, depending on submerging the plant roots in a deep nutrient solution, supplies oxygen *via* an air pump	• High biomass yield • Low initial cost and reliability • No nutrient pump needed	• Slow growth rate • Common root diseases, such as root rot due to a lack of dissolved oxygen • The nutrient supply tank must be refilled regularly	[29]
Ebb and flood system (EFS)	It works by periodically flooding plant roots in a growing channel with nutrient solution	• Recirculating facility • Seed germination equality • Low maintenance required	• Disposed to algae growth • Clogging of recirculation valves • Any damage to pumps or valves leads to crop loss	[30]
wick watering system (WWS)	It is established by settling plants in porous growing media, which are immersed in liquid *via* capillary action	• Low maintenance • Perfect for smaller plants • Simple and easily affordable	• No recirculation facility • Prone to algae growth • Limited oxygen excess	[31]
Drip hydroponics system (DHS)	Relying on a drip irrigation mechanism to supply nutrient solution individually to plant roots	• Extreme nutrient absorption • Sufficient oxygen excess • Recirculation ability	• Clogging of holes • Regular cleaning is required • Disposed to algae growth.	[29]

Soilless system	Definition	Merits	Demerits	References
Aeroponics system	This system works by hanging plant roots in a dark chamber in the air and exposing them to fog or nutrient solutions sprayed out of nozzles	• Plentiful oxygen supply • Faster plant growth and root expansion • Maximize nutrient absorption • water savings reach 90–95%	• Clogging of nozzles • little suited for organic-based nutrients such as digestate • Need filtration, besides high energy consumption • High production cost	[32, 33]
Aquaponics system	A combined production system gathering aquatic organisms (fish) and a plant growth system (hydroponics) supplied with nutrients derived from waste originating from feeding the aquatic organisms	• Dual productivity system, besides its sustainability and eco-friendly approach *via* valorization of fish waste • Support organic farming by using organic waste as nutrients, where no synthetic nutrient is required	• Increased initial cost to construct. • Require more space due to the dual production system of fish and plants • Need regular monitoring for some parameters such as ammonia nitrate, pH, and EC • Disease management is required due to - loop system • Limited crop selection for growth • High energy consumption	[34]

Table 1.
Comparative evaluation of major soilless cultivation systems. Technical specifications, advantages, limitations, and performance metrics.

2.2 Aeroponics

Aeroponics is considered an advanced and improved method for hydroponics. Aeroponics is a method of growing plants in an air or fog environment without the utilization of soil. In this system, plant roots are hanging in the fog air and intermittently sprayed with a nutrient solution [33]. The major difference between aeroponics and hydroponics is that the solution of nutrients ([32] #63) is sprayed as a fine drop or mist in the root growth zone. This leads to rising and enhanced oxygen access to the crops' roots for greater absorption, promoting faster growth and higher yields. Additionally, only sponges or net pots are used to hang and support the plant roots. Aeroponics is often used in controlled environment agriculture, such as vertical farms and greenhouses, and is considered an efficient and sustainable approach to plant cultivation [32].

Crops type	Crops name
Cereals	*Oryza sativa* (Rice), *Zea mays* (Maize)
Fruits	*Fragaria ananassa* (Strawberry)
Vegetables	*Lycopersicon esculentum* (Tomato), *Capsicum frutescens* (Chili), *Solanum melongena* (Brinjal), *Phaseolus vulgaris* (Green bean), *Beta vulgaris* (Beet), *Psophocarpus tetragonolobus* (Winged bean), *Capsicum annum* (Bell pepper), *Brassica oleracea* var. *capitata* (Cabbage), *Brassica oleracea* var. *botrytis* (Cauliflower), *Cucumis sativus* (Cucumbers), *Cucumis melo* (Melons), *Raphanus sativus* (Radish), *Allium cepa* (Onion)
Leafy vegetables	*Lactuca sativa* (Lettuce), *Ipomoea aquatica* (Kang Kong)
Condiments	*Petroselinum crispum* (Parsley), *Mentha spicata* (Mint), *Ocimum basilicum* (Sweet basil), *Origanum vulgare* (Oregano)
Flower/ornamental crops	*Tagetes patula* (Marigold), *Rosa berberifolia* (Roses), *Dianthus caryophyllus* (Carnations), *Chrysanthemum indicum* (Chrysanthemum)
Medicinal crops	*Aloe vera* (Indian Aloe), *Solenostemon scutellarioides* (Coleus)
Fodder crops	*Sorghum bicolor* (Sorghum), *Medicago sativa* (Alphalfa), *Hordeum vulgare* (Barley), *Cynodon dactylon* (Bermuda grass), *Axonopus compressus* (Carpet grass)

This table catalogs 32 crop species successfully cultivated in soilless systems, organized by type (cereals, vegetables, medicinal plants, etc.). Each entry includes the Latin binomial name to ensure precise identification. Highlighted crops include high-value varieties like tomatoes (Lycopersicon esculentum) and strawberries (Fragaria ananassa), as well as niche species such as medicinal Aloe vera and ornamental Chrysanthemum indicum. This demonstrates the versatility of soilless methods across agricultural sectors.

Table 2.
Crop suitability for soilless cultivation: comprehensive list of species grouped by category with scientific Names [35].

2.3 Aquaponics

Aquaponics is a hybrid of hydroponics and aquaculture in which plants grown in a soilless culture, frequently under regulated conditions, use nutrients provided by growing fish. Comparing aquaponics to traditional aquaculture, some potential benefits include increased sustainability, lower resource usage, and less environmental impact [36].

The ultimate objective is to increase sustainability by using the fish culture system's nutrient waste as a source of nutrients for plant growth, producing two kinds of commercially viable goods. The idea is that aquaculture and plant culture combined into one system will work well enough to be profitable. Both systems will be subject to limits, and management complexity will rise because of the integration. The fish feces from the feed intake give the plants vital nutrients. Ammonia, nitrites, nitrates, phosphates, potassium, and organic carbon are all eliminated from the water by the plants and can subsequently be given back to the fish, as shown in **Figure 1**. Numerous surface areas offered by root systems allow bacteria to extract biological oxygen demand (BOD) from the water [34]. This method is green and sustainable because it uses less fertilizer, does not need pesticides, and has a high water usage efficiency. Since aquaponics is becoming more and more popular, the main obstacle is making it viable and dependable on a commercial scale [37]. Furthermore, it is considered a sustainable approach to the circular economy, where the cultivated crops depend on the fish feces to supply plants with nutrients, especially with the presence of bacteria that convert the noxious fish wastes, such as ammonia, into nitrates, which are suitable for plant growth [38].

3. Technological innovations and cutting-edge soilless systems

Soilless systems based on modern technology and automation are rapidly advancing, offering promising solutions for sustainable agriculture nowadays and beyond. Smart soilless culture systems driven by artificial intelligence(AI), Internet of Things (IoT), and machine learning (ML) are the most advanced technology of these developments, enabling real-time monitoring and precise control of environmental parameters such as temperature, humidity, pH, and nutrient levels [39]. These technologies aim to optimize plant growth conditions while minimizing dependence on resources, which leads to increased efficiency and productivity of crops. Especially when integrating soilless system-based real-time monitoring and control with renewable sources of energy, such as solar and wind energy, as shown in **Figure 3**. Where the integration of AI and ML is supporting the growth control systems, which is revolutionizing indoor growth, allowing for individualized plant care and early detection of growth anomalies [40].

Furthermore, modern designs of soilless techniques, such as vertical farming, maximize land use efficiency in urban environments. Besides, progressive LED lighting systems with dynamic light spectra enhance photosynthesis and plant development while reducing energy consumption. These cutting-edge technologies are not only improving crop yields but also addressing critical challenges in water conservation, with some soilless systems using up to 90–95% less water than traditional cultivation methods [41]. As these innovations continue to evolve, they promise to transform agricultural practices, making food production more sustainable, efficient, and adaptable to the challenges of climate change and urbanization.

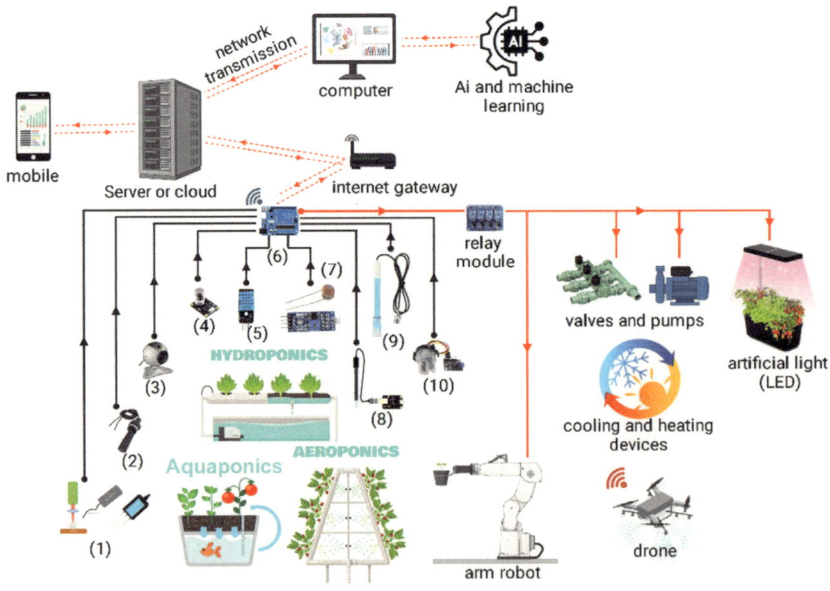

Figure 3.
Advanced monitoring and sensing technology-based AI applied for soilless cultivations toward optimization and full automation: (1) nutrient concentration sensor, (2) water level sensor, (3) camera, (4) CO_2 level sensor, (5) temperature and humidity sensor, (6) microcontroller-base WiFi, (7) light sensor (LDR), (8) electrical conductivity (EC) sensor, (9) pH sensor, and (10) turbidity sensor.

3.1 Data acquisition and IoT

Data acquisition (DAQ) is the process of collecting, measuring, and converting physical phenomena into digital data that can be analyzed and processed by computers, as shown in **Figure 3**. It involves sampling signals from sensors or transducers that measure physical conditions such as temperature, pressure, or electrical signals and then converting these analog signals into digital values [42]. Nowadays, DAQ systems have become a fundamental requirement in modern soilless culture techniques, which enable precise monitoring and control of environmental parameters critical for plant growth. These systems typically incorporate a network of sensors to measure factors such as temperature, humidity, pH, electrical conductivity (EC), dissolved oxygen (DO), and nutrient concentrations in real time [43]. IoT technologies are gradually being integrated into these systems, allowing for remote monitoring and automated control of cultivation conditions [44, 45]. The Internet of Things (IoT) refers to a network of interconnected physical objects, devices, structures, and advanced sensing technologies, such as Radio Frequency Identification (RFID) and various sensors. By integrating the natural environment with digital computing systems and online platforms, IoT generates critical data and enhances functionality, ultimately driving improved productivity and efficiency [46]. For instance, in aeroponic systems, sensors can monitor mist delivery intervals and droplet size, which is vital for preserving optimal root zone moisture. The integration of these technologies, besides enhancing crop yield and quality, significantly improves resource use efficiency, particularly in water and nutrient management [47]. As these technologies continue to develop, they are paving the way for more effective and sustainable cultivation methods, particularly in urban and controlled environmental agricultural settings.

3.2 Role of machine learning and artificial intelligence in soilless culture techniques

Machine learning and artificial intelligence are revolutionizing soilless culture techniques by enhancing precision, efficiency, and productivity. These advanced technologies enable real-time monitoring and control of crucial environmental parameters, optimizing plant growth conditions and resource utilization. AI-driven systems can analyze vast amounts of sensor data to make predictive decisions about nutrient delivery, climate control, and pest management. Rahman and Chakraborty [45] employed deep learning models for automated plant phenotyping, including species recognition and health condition assessment, allowing for early detection of plant stress and disease. Machine learning algorithms are also being used to optimize nutrient delivery profiles tailored to specific crop varieties, leading to improved yields and resource efficiency [48]. Mamatha and Kavitha [49] created an AI system integrated with aeroponic systems to monitor and control mist delivery intervals and droplet size, which is crucial for maintaining optimal root zone moisture. Furthermore, the integration of AI with IoT technologies has facilitated the development of smart, automated growing environments that can be remotely monitored and controlled, which significantly reduces labor costs and improves overall system management [50].

4. Water efficiency and conservation

Conventional cultivation is reliant on freshwater as a major source of irrigation, which plays a pivotal role in crop production. However, irrigation practices often lead to crucial water inefficiencies and environmentally harmful waste [51]. The conventional agriculture sector is considered the sector that consumes fresh water, consuming approximately 70% of global freshwater [52]. The consumption of water is mostly applied *via* irrigation networks or rainfall harvesting. The crops' accurate irrigation water requirements rely on specific climatic variables, soil conditions, and plant factors [53]. Generally, the water requirements for cereal grains are about 300–500 liters of water per kg output, whereas the water requirements for vegetables are 300–700 liters/kg [54]. Conventional irrigation methods, which are applied to traditional agriculture, led to enormous water inefficiency due to water loss. For instance, surface irrigation methods such as flood and furrow irrigation systems have low efficiency in distributing the water, which results in overwatering in some areas and not enough moisture in other areas, which leads to asymmetric growth that affects plant productivity [51]. Sprinkler irrigation systems suffer from irregular spray due to spray drops drifting and evaporation losses, which may reach 50% of water delivered from the source [55]. Irrigation water losses in conventional cultivation may be due to widespread percolation to groundwater or runoff. The irrigation efficiency in the developing countries is about 30–50%, while in the developed countries it may reach above 60% [56]. Recurring overwatering leaches the nutrients from the soil to the groundwater, participating in salinity buildup and erosion [57].

On the other hand, soilless culture techniques, or hydroponics, require significantly less water than soil-based cultivation. Where hydroponics uses 10 times less water than traditional cultivation to produce the same yield [58]. For instance, producing 1 kg of tomatoes needs 20–40 liters of water in hydroponics compared to soil cultivation, which takes 214 liters [59]. There are several reasons to make hydroponics save more water and make plants thrive on a small portion of the water, where no water is lost due to soil deep percolation or evaporation. Furthermore, hydroponics recirculates the nutrient solution, and accurate moisture control helps to avoid overwatering. Besides, real-time monitoring of the moisture level and irrigation intervals is required. In soilless culture, higher yields are achieved per liter of water used (high water use efficiency). In addition, hydroponics optimizes the growth conditions surrounding the plant. The circulation of the nutrient solution reduces the footprint compared to soil cultivation *via* using a lower amount of mineral nutrients or depending on organic nutrient solutions such as biogas digestate [60].

Furthermore, the recirculation of the nutrients makes discharge negligible, while soil cultivation releases contaminated effluents [5]. Hydroponic systems demonstrate enhanced nutrient uptake efficiency under optimal conditions, effectively preventing nutrient surplus. Besides conserving water, it mitigates the risk of eutrophication in adjacent ecosystems, which can lead to harmful algal blooms and the formation of dead zones.

Adopting vertical cultivation practices allows for the strategic placement of production facilities near urban areas, which significantly reduces transportation distances and minimizes spoilage losses. Additionally, the use of energy-efficient LEDs, specifically designed to optimize photosynthesis, reduces electricity consumption, while the incorporation of renewable energy sources further diminishes the overall environmental impact [61].

5. Controlled environmental agriculture (CEA)

Controlled environment agriculture (CEA) is an advanced farming method that involves growing crops in controlled environments, such as greenhouses, vertical farms, and indoor facilities [62]. CEA allows precise regulation of factors like temperature, humidity, light, and nutrient levels, enabling the best plant growth regardless of external conditions. Adopting vertical cultivation practices allows for the strategic placement of production facilities near urban areas, which significantly reduces transportation distances and minimizes spoilage losses. CEA decreases the danger of soil-borne diseases and pests while conserving water and nutrients through recirculating systems [17]. CEA is considered a green solution for urban cultivation and regions with limited arable land.

One of the principal roles of CEA in soilless farming is its ability to maximize resource efficiency. Additionally, it enables year-round production, overcoming seasonal limitations and ensuring a stable supply of fresh produce. This is significant specifically in referring to food security challenges in urban areas and regions with harsh climates [63]. Furthermore, CEA allows for the farming of high-value crops such as leafy greens, herbs, tomatoes, and strawberries with enhanced quality and nutritional value [64]. CEA can optimize photosynthesis and improve crop yield and quality (taste, color, and shelf life) by controlling light spectra using LED technology [65].

However, CEA has several challenges despite its benefits. It entails high initial capital costs to invest in infrastructures like climate control systems and LED lighting [66]. Another key concern includes energy consumption since maintaining optimal conditions in the growth cycle usually requires large amounts of electricity, especially for lighting and heating. However, a modern approach to integrating renewable energy and energy-saving technologies can present avenues to surmount these issues. For instance, CEA is increasingly using solar-powered greenhouses and energy-saving LEDs to reduce its carbon footprint [67]. Additionally, the incorporation of automation and artificial intelligence in CEA systems can enhance precision farming, reduce labor costs, and improve crop management [68].

6. Low-quality water (unconventional resources)

6.1 Seawater application in soilless culture

Sea or saline water is considered the most plentiful water source on the planet and has a semi-balanced ionic environment for plants in its specific composition [69]. Despite its high chlorine content (about 75, 10% of NaCl and $MgCl_2$, respectively). Seawater is rich in most of the nutrients that plants require, as shown in **Table 3**, besides containing the necessary trace elements and microorganisms [71]. Several studies revealed that the planet exposed to seawater had a plant salinity tolerance that improved twice compared to NaCl at the same level of EC in the majority of species [71]. Even while the findings of an experiment on red leaf lettuce point to a similar conclusion, such a claim has not yet been sufficiently substantiated by more specialized research. Breckle [77] states that large-scale sustainable agriculture that depends on pure seawater is not practical for several reasons, including the sodium content of full-strength seawater, which causes toxic effects for most species and degrades soil structure, particularly in the absence of an effective leaching system. However, the goal is to restore the lost coastal soil while cutting down on freshwater inputs. On the other hand, small-scale seawater irrigation systems may be appropriate and economically practical, especially for coastal areas that have plenty of saline water. In a recent study developed by Yuan Longping and his group, they developed a new strain of rice producing a higher yield compared to most commercial rice, which is grown at a salinity of 10% seawater [71]. Furthermore, de Vos and Bruning [78] studied some

Characteristics and elements	Hoagland [70]	Seawater [71, 72]	Digestate [73, 74]	Urine [75, 76]
pH	6–6.5	7.5–8.4	5.8–13.7	6.0–7.5
TDS (ppm)	1100–1300	35,000-45,000	1632	5120–12,800
TN	210	0.9	17.65(g/kg)	2000–4000 (mg/L)
P	31	0.1	5.8–30 (g/kg)	200–800(mg/L)
K	235	380	1.2–10(g/kg)	1000–2000(mg/L)
Mg	48	1272	0.1–14.1(g/kg)	20–100(mg/L)
Ca	200	400	0.6–98.5(g/kg)	50–150(mg/L)
S	64	884	11–17(g/kg)	100–300(mg/L)
Mn	0.5	0.01	ND	ND
B	0.5	4.6	ND	ND
Fe	1 up 5	0.02	24–79(g/kg)	ND
Zn	0.05	.014	238–5923(mg/kg)	ND
Cu	0.02	0.09	ND	ND
Mo	0.01	0.002	ND	ND
Cl	ND	18,980–20,000	ND	1000–3000(mg/L)
Na	ND	10,561	ND	1000–3000(mg/L)
Organic matter	ND	ND	15.6–98.0%	ND
C/N ratio	ND	ND	7.9–8.2	ND

[1]*All element concentrations are in ppm unless another unit.*
[2]*Digestate variable depends on feedstock; the data mentioned relates to sewage sludge (Composted Industrial and Municipal Water Waste).*
[3]*Concentration values g/kg or mg/kg calculated based on a dry mass.*
[4]*ND refers to not determined.*

Table 3.
Chemical composition analysis. Unconventional nutrient sources vs. standard Hoagland solution for soilless cultivation.

plant species (potato, cabbage, onion, carrot, lettuce, and barley) salinity tolerance by exposing them to seawater. The result revealed that some types of experimental crops could be grown without any yield loss under a saline water level of 5–7 $dS\,m^{-1}$. Despite several drawbacks, seawater as an additional irrigation supply appears to address the demand for resource-efficient and sustainable agricultural methods. Furthermore, irrigation with seawater leads to contamination by salts, especially Ca^{2+}/Na^{+} [79]. The ability to apply seawater as a solution to replace the water in the soilless systems is principally represented by the highest precise control over the water supply. Seawater, as a complementary source for soilless culture techniques, may achieve several goals, such as reducing the consumption of freshwater by replacing a part of it, utilizing the seawater's nutrient content, and improving the plant metabolites in certain cases. However, the main constraints related to the reduction in crop yield as a result of salinity stress are also sodium and chloride levels rising in the edible products [80].

6.2 Biogas digestate applications in soilless culture techniques

Biogas digestate (anaerobic digestate) is a nutrient-rich byproduct of anaerobic digestion for organic waste such as animal, food, and agricultural wastes [81]. Anaerobic digestion involves the breakdown of organic matter by microbes in oxygen-absent conditions, producing biogas and digestate as primary outputs [82]. Recently, digestate has gained attention for its potential as a sustainable fertilizer in soilless culture techniques such as hydroponics and aeroponics. It is considered one of the promising digestate valorization strategies for perfect integration between food production and organic waste management [83]. Digestate is rich in nutrients like nitrogen, phosphorus, and potassium, making it an attractive alternative to synthetic fertilizers [74]. In soilless cultivation systems, digestate can be used as a sole nutrient source, offering several environmental benefits. Its application in soilless cultivation systems aligns with the principles of circular economy and sustainable agriculture as it repurposes waste into a resource, reducing dependency on fossil fuels and chemical fertilizers, minimizing environmental pollution, and conserving water [84]. Moreover, the production of traditional synthetic fertilizers is energy-intensive and frequently leads to water pollution and eutrophication. In contrast, digestate provides a renewable and eco-friendly nutrient source, reducing greenhouse gas emissions associated with fertilizer production and waste disposal. Moreover, its application in closed-loop systems like aeroponics and hydroponics enhances nutrient use efficiency. Also, these systems allow precise control over nutrient delivery, minimizing losses and maximizing plant uptake [85]. This approach contributes to mitigating climate change by recycling organic waste and reducing the carbon footprint of agriculture, besides supporting sustainable crop production, as shown in **Figure 4**, where the application of digestate in soilless techniques aligns with many United Nations sustainability goals.

Several studies have demonstrated the potential of biogas digestate in soilless cultivation. For instance, Möller and Müller [86] emphasized the suitability of digestate as a nutrient source in hydroponic systems, revealing that it can support plant growth while maintaining nutrient balance. Similarly, a study by Lai and Tu [87] found that digestate-based liquid fertilizers performed comparably to conventional fertilizers in hydroponic lettuce production, above the benefit of reducing environmental impacts. Furthermore, the integration of digestate and aeroponic systems has shown a promising approach to optimizing resource utilization, besides the fine misting of nutrient solutions in aeroponics, ensuring effective nutrient absorption by plant roots [5]. In addition, Malik et al. [88], and Weimers et al. [89] indicate in their studies that digestate can enhance plant growth and productivity by providing essential nutrients, although its effectiveness may vary depending on factors such as nutrient availability and pH management. However, challenges such as ammonia toxicity and nutrient imbalances must be addressed through the application of digestate as a source of nutrients to optimize digestate nutrient availability and uptake by plants [90].

By converting organic waste into a valuable resource (biogas and digestate), this approach reduces reliance on synthetic fertilizers, besides addressing critical issues such as waste management and climate change. To ensure compatibility with diverse soilless systems and crop types, further research is necessary to optimize digestate pretreatment and application methods. As the global demand for sustainable food production grows, biogas digestate could play a pivotal role in forming the future of agriculture and fostering a more resilient and environmentally friendly food system.

6.3 Urine in soilless culture techniques

The utilization of human urine in soilless culture systems, often referred to as 'anthroponics', has gained attention as a sustainable approach to nutrient exploitation in soilless agriculture. This method leverages the urine nutrient content, particularly nitrogen, phosphorus, and potassium [91].

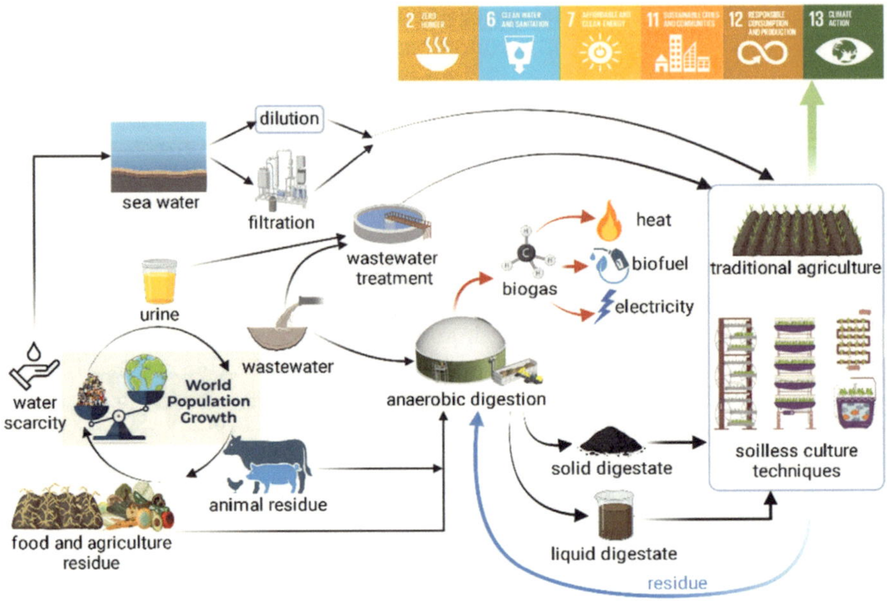

Figure 4.
Biocircular economy model for sustainable agriculture: From waste valorization to soilless cultivation. The figure illustrates that increased world population leads to increased demand for food and water, hence increasing waste such as food and agricultural residue, animal residue, wastewater, etc., which requires non-conventional solutions for waste management such as treatment or anaerobic digestion and nutrient recovery (energy and fertilizers). These products support both soilless and traditional agriculture and advance multiple UN Sustainable Development Goals (SDGs).

Many studies demonstrated the feasibility and potential benefits of incorporating urine-derived nutrients into soilless cultivation. Such as Bliedung et al. [92] successfully used nitrified synthetic urine to grow lettuce in a hydroponic system; the system achieved comparable yields to conventional nutrient solutions. However, the nitrification process is essential in urine treatment before use, as it reduces the ammonia level and preserves other plant nutrients. In addition, challenges like high salt content and potential pathogen presence necessitate careful pretreatment and dilution of urine before application. Volpin et al. [93] addressed these problems by using membrane distillation to concentrate urine and then diluting it for hydroponic use. The optimal dilution ratio varies depending on the crop and system design, ranging from 1:10 to 1:50, and can be applied in hydroponics. Another study revealed that lettuce grown in fully nitrified urine concentrate outperformed those grown in partially nitrified urine, highlighting the importance of proper urine processing [94]. On the ecological level and beyond crops, the use of urine in hydroponics offers significant environmental benefits, including reduced reliance on synthetic fertilizers and improved nutrient recycling. Magwaza et al. [15] demonstrated that the partially treated domestic wastewater, which includes urine, could serve as a viable nutrient source for hydroponic tomato production when combined with commercial fertilizers.

On the other hand, some studies using soilless culture techniques as a treatment step for urine besides the crop production, such as Yang et al. [91] study, which aimed to assess the effectiveness of a hydroponic system using spinach to treat urine after partial nutrient recovery to meet sewer discharge standards. The results showed that plants grown in urine diluted at a 1:50 ratio exhibited similar growth traits (such as growth rate and leaf number) to those grown in a standard nutrient solution.

Additionally, the system effectively reduced chemical oxygen demand (COD), total nitrogen (TN), and total suspended solids (TSS) levels, with the treated effluent complying with local sewer discharge regulations.

This approach tackles nutrient management and supports water conservation initiatives [15]. However, despite the encouraging outcomes, several challenges persist in scaling up urine-based hydroponic systems. These include overcoming public perception issues, navigating regulatory barriers, and developing efficient infrastructure for urine collection and treatment for further applications.

6.4 Utilization of wastewater in soilless culture techniques

Because wastewater contains nutrients like N (64 mg/L) and P (6.2 mg/L), it is perfect for utilizing to produce leafy crops [95] as it minimizes the amount of minerals applied by farmers [96]. Due to the higher concentration of nutrients in raw wastewater (K^+ = 18 mg/L, Mg^{2+} = 22 mg/L, Ca^{2+} = 53 mg/L, NO^{3-} = 6.5 mg/L, NH^{4+} = 4.3 mg/L, P = 8 mg/L, SO_4^- = 84.8 mg/L), leafy crops (lettuces, celery, cauliflower, and broccoli) supplied with raw wastewater grew better than those irrigated with aquifer water [95]. Several studies highlighted the potential of wastewater application in soilless culture techniques, especially hydroponics [97–99]. They mentioned that the application of wastewater in hydroponics achieves additional income when utilized for economic crop cultivation, besides the biological treatment of the wastewater fraction. Furthermore, **Table 4** mentions some studies on the successful application of wastewater in soilless culture systems and hydroponics.

Although using treated wastewater in hydroponic systems to grow vegetables has been shown to have several advantages, the adoption of these systems may be restricted due to the decreased crop quality and shelf life. Magwaza et al. [109] and Carvalho et al. [102] caution that even while wastewater contains both macro- and micronutrients, little is known about how to add them properly, which can impede plant growth by limiting or oversupplying nutrients.

Most of the applications of wastewater in hydroponics aim to treat wastewater as the main target, and then crop production. For instance, in an experiment that established a hydroponic system for cultivating lettuce with wastewater, the system accomplished the removal of 80% of nitrogen, 77% of phosphorus, an 86% reduction in COD, 87% reduction in biological oxygen demand (BOD), and 99% of TSS. Where the initial concentration of the recovered wastewater was 64 mg/L of nitrogen, 6.2 mg/L of phosphorus, 195 mg/L BOD, 366 mg/L COD, and 150 mg/L TSS [110]. Vaillant and Monnet [111] achieved comparable disposal levels, where they removed 98, 91, and 82% of TSS, BOD, and COD, respectively. They ascribed the higher elimination to the elements' mineralization and degradation, as well as the filtration process carried out in the toloache (*Datura ferox*) roots. The wastewater's high P load likely accounted for the elimination of 96% of the nitrogen and 38% of the P.

Furthermore, in an experiment conducted by Yang and Zheng [112], where they changed hydraulic retention durations to 3, 2, and 1 day, respectively, and achieved removals of NO_3, NO_2, and N_2 by 91, 97, and 71%. They employed a floating system that encouraged contact between roots and microorganisms required for the nitrification operation and favored denitrification by lowering the concentration of dissolved oxygen.

However, the usage of raw wastewater can have certain drawbacks because of its composition, such as the existence of pathogens, such as bacteria, viruses, protozoans, and helminths [109]. For this reason, strict laws enforced by the European Union ensure that the application of raw wastewater for crop irrigation poses little harm to human health.

Wastewater type	Soilless culture technique	Plant type	References
Urban wastewater	The deep flow technique achieved nitrogen and phosphorus removal	Lettuce	[92]
Domestic rinse water	Indoor hydroponic system proposed as a novel model for urban households	Vegetables and flowers	[100]
Domestic wastewater	Partially–hydroponic	Sunflower	[98]
Tertiary-treated wastewater	Auto-pot watering hydroponic system	Wheat (ESWYT and SAWYT varieties)	[101]
Domestic wastewater	Nutrient film technique	Lettuce, gerbera	[102, 103] [104]
Reclaimed urban wastewater	Hydroponic chambers	Barley forage	[105]
Partially treated domestic wastewater	Deep flow technique	Tomato	[15]
Anaerobic digestate	Media bed	Silverbeet	[106]
Domestic wastewater	Substrate-bed based	Winter squash, green beans, sweet corn, eggplants, cherry tomatoes, various cutting flowers, Citrus, olives, and rosemary	[107]
Fishpond wastewater	Hydroponic reactor	Lettuce	[108]

The table synthesizes 10 international studies demonstrating wastewater applications in hydroponics. Parameters include system types (e.g., NFT and DFT) and crop types that are used in these systems.

Table 4.
Diverse successful studies for the application and reclamation of wastewater in soilless culture techniques.

Based on the degree of treatment it undergoes, reclaimed water is divided into four classifications. Class A is designated for all crops, including those consumed fresh, even if the edible part has come into direct contact with the raw wastewater. Classes B and C apply to crops that are processed and not intended for human food, which are permitted for raw crops as long as the edible portion is kept out of direct contact with wastewater. Class D is utilized for crops for seed production purposes. Each class has definite water quality demands; for instance, Class A requires improved sterilization and $\leq$ 10 CFU/100 ml of *Escherichia* **coli**, BOD $\leq$10 mg/L, and turbidity $\leq$10 NTU [113].

Additionally, recommendations for the extreme concentrations of hazardous elements in reclaimed wastewater employed for agricultural irrigation have been established by the FAO and WHO: Al and Pb 5.0 mg/L, Li_2 2.5 mg/L, Zn 2 mg/L, F 1 mg/L, Cu, Mn, and Ni 0.2 mg/L, Co 0.05 mg/L, As, Be, Cd, and Cr 0.1 mg/L, Mo and Se 0.01 [114].

7. Energy requirements and costs

Hydroponic and aeroponic systems are soilless growing techniques that rely largely on electrical equipment to function; hence, energy consumption is an important factor to consider. These systems might have significant energy requirements, even though they have several benefits in terms of agricultural productivity, water efficiency, and

DOI: http://dx.doi.org/10.5772/intechopen.1010669

year-round production. Because hydroponic systems require pumps, lighting, and climate control systems, they usually use more energy than traditional soil-based agriculture [115]. A small hydroponic system can use between 200 and 500 watts of electricity per day, with larger systems potentially requiring up to 1500 watts or more [116]. Although aeroponic systems typically use less water than hydroponic ones, they may initially cost more because high-pressure pumps are needed to produce the tiny mist needed for root nutrition [115]. The energy consumption in these systems varies significantly based on factors such as system design, crop type, and environmental conditions. Indoor hydroponic and aeroponic setups increasingly use LED lighting, which has significantly reduced energy consumption compared to traditional high-intensity discharge (HID) lamps [117]. However, especially in completely regulated conditions, lighting still accounts for a significant amount of the energy budget. Regmi and Rueda-Kunz [118] revealed that vertical farming systems require more illumination at various levels and frequently use hydroponic or aeroponic techniques; they can use more energy than horizontal systems. Despite the higher energy demands, these systems can be more energy-efficient when considering productivity per unit of energy input. A study comparing hydroponic and aeroponic lettuce production to conventional soil-based methods found that while the soilless systems consumed more energy, they produced significantly higher yields, potentially offsetting the increased energy costs [119]. Moreover, precise control over growing conditions in hydroponic and aeroponic systems leads to faster growth rates and shorter cultivation cycles, which may result in lower cumulative energy consumption per crop cycle [47]. Ongoing efforts aim to enhance energy efficiency in these systems. Strategies involve optimizing lighting schedules, using energy-efficient pumps and climate control systems, and incorporating renewable energy sources. Some researchers have investigated the potential of solar or wind power to reduce the energy requirements of hydroponic and aeroponic systems, which could lead to greater sustainability over time [43]. Additionally, advancements in automation and smart sensing technologies are enabling more precise control over system operations, potentially reducing redundant energy expenditure [120]. As the agricultural sector encounters escalating demands to mitigate its environmental impact, the energy consumption of hydroponic and aeroponic systems emerges as a critical focus for research and development. Although these systems present viable solutions for sustainable food production, especially in urban settings, their energy requirements necessitate meticulous management to ensure long-term viability and environmental advantages. Future research should prioritize the optimization of energy utilization within these systems, the exploration of innovative energy sources, and the execution of comprehensive life cycle assessments to fully elucidate their energy footprint relative to conventional agricultural practices.

8. Economic analysis: Expense-benefit considerations

The economic feasibility and potential for broad adoption of soilless growing methods, especially hydroponics and aeroponics, are greatly impacted by the intricate interactions between costs and benefits that are revealed by economic analysis. Compared to conventional soil-based agriculture, these modern cultivation techniques provide several benefits, such as improved crop yields, year-round production potential, and resource efficiency [121]. They do, however, come with particular expenses and difficulties that need to be carefully evaluated.

The initial costs for setting up hydroponic and aeroponic systems are usually higher than those for traditional farming methods. This is essentially because they need particular equipment and infrastructure [122]. These costs include growing structures, irrigation systems, lighting, and climate control technologies. The initial costs can be substantial, potentially acting as a barrier to entry for small-scale farmers or those in resource-limited settings [122]. Despite higher initial costs, operational expenses for hydroponic and aeroponic systems can be lower in certain aspects. Resh [5] demonstrates that hydroponic farming can reduce water usage by up to 90% compared to conventional methods, leading to significant cost savings in water-scarce regions. Similarly, Savvas and Gruda [123] highlighted the precise nutrient delivery in these systems, which minimizes waste and optimizes plant growth, potentially reducing fertilizer costs.

The economic advantages of soilless farming often come from its ability to boost productivity and make better use of resources. Some studies consistently demonstrate that hydroponic and aeroponic systems produce higher crop yields than traditional farming, such as a study on commercial hydroponics in Nigeria that found positive net present values and impressive internal rates of return around 83% for both small- and medium-scale operations over 10 years [10]. This evidence suggests these systems are not only economically viable but also offer significant profit potential.

9. Sustainability considerations

Soilless cultivation systems, including hydroponics, aeroponics, and aquaponics, are considered promising solutions for sustainable agriculture in the face of global challenges such as climate change, water scarcity, and food security. These systems offer significant advantages over traditional soil cultivation, including higher efficiency for resource utilization, reduced environmental impact *via* reducing synthetic nutrients as a result of recirculation of nutrients inside these systems, besides depending on non-traditional organic sources for nutrition, and increased crop yields [60, 124, 125]. However, the long-term sustainability of soilless systems requires careful consideration of various factors, including energy consumption, waste management, and economic viability [126].

Furthermore, the environmental impact assessment of soilless culture techniques reveals a composite image of their ecological footprint. While these systems generally use less water and land than conventional agriculture, they may have higher energy requirements, especially for indoor operations [41]. For instance, greenhouse hydroponics can reduce up to 90% of water usage compared to traditional cultivation, but the energy needed for lighting and climate control can contribute to a larger carbon footprint [126]. This dilemma can be addressed by drawing attention to the application of renewable energy, including solar energy, wind energy, and alternative options such as biogas and organic hydrogen [27].

Gebreegziher [127], in his life cycle assessment (LCA) study, demonstrated that the environmental impact of soilless systems varies depending on factors such as the energy source, cultivation method, and crop type. The long-term sustainability of soilless systems relies on several major factors, such as the integration of renewable energy sources, like solar or wind power, which can significantly reduce the carbon footprint of these operations [27, 128]. Also, the development of more efficient lighting technologies, such as LED systems, can lower energy consumption while maintaining crop yields [129]. On the other hand, the implementation of circular economy principles, including the recycling of nutrients and growing media, can enhance resource efficiency and reduce waste [130]. Additionally, the use of biodegradable materials for plant containers and the adoption of integrated pest management techniques can further improve the environmental performance of soilless systems [131].

Soilless culture techniques align with several United Nations Sustainable Development Goals (SDGs), particularly SDG 2 (zero hunger), SDG 6 (clean water and sanitation), SDG 11 (sustainable cities and communities), and SDG 13 (climate action) [132–135]. Soilless technologies help provide food security and lessen the need for long-distance produce transportation by allowing year-round production in metropolitan areas and areas with little arable land. Above all, the high water use efficiency in soilless systems solves the issue of water scarcity, while their potential for integration into urban environments supports the development of sustainable cities [136, 137].

Furthermore, there are many strategies for enhancing the sustainability of soilless techniques, such as optimizing resource utilization *via* precision agriculture and smart monitoring systems [8]. Developing and implementing energy-efficient technologies, like advanced climate control systems and energy-saving lighting, is also a

Case study	Description	References
Singapore's urban farming initiative	Considered extensive urban farming projects, including hydroponics and aeroponics, to increase local food production. The city-state aims to produce 30% of its nutritional requirements locally by 2030.	[136]
The Plant Chicago	This project serves as a model for urban vertical farming. It combines aquaponics and hydroponics to revitalize an abandoned industrial site, fostering sustainable economic development and providing fresh yield to local communities.	[140]
Hydroponic green farming initiative (HGFI) in Jordan	This initiative promotes hydroponics as a solution to water scarcity, enhancing economic opportunities and providing vocational training to underserved communities.	[141]
Wageningen University's hydroponic systems	The Netherlands has been at the forefront of hydroponic research; its hydroponics projects revealed its potential to achieve higher yields while using significantly less water.	[142]
Commercial-scale hydroponics in Nigeria	The hydroponic cultivation of leafy and fruity vegetables in southwest Nigeria shows substantial economic viability for potential small-scale and medium-scale farmers. The cost-benefit analysis of the production scales shows that they are economically feasible in the short and long terms.	[10]
Greenhouse hydroponics in Skidmore	These projects support alternative farming production methods, specifically hydroponics, with considerable benefits academically, sustainably, and even economically.	[143]
Project management plan for Belize City	This project plan is set to effectively and efficiently manage the construction of an urban hydroponics farm within the residential community in the Kings Park Area of Belize City, Belize. Given the challenges of sustainability and food security facing the world, including Belize, in the face of growing urbanization	[144]

This table documents seven pioneering soilless agriculture projects worldwide, analyzing their technical approaches, resource efficiencies, and socio-economic benefits.

Table 5.
Global implementations of commercial soilless farming. Project profiles, sustainability outcomes, and economic impacts.

key strategy. Besides exploring alternative growing media, including recycled materials and biodegradable substrates, to reduce reliance on non-renewable resources [126, 138]. Moreover, integrating soilless systems with other sustainable practices, such as rainwater harvesting and composting, and using non-conventional nutrient sources, such as biogas digestate and wastewater, to create closed-loop production cycles [128]. Also, research should be invested to improve the efficiency and productivity of soilless systems while minimizing their environmental impact. Additionally, efforts are being made to promote training programs that aim to build capacity in sustainable soilless cultivation techniques, besides developing supportive policies and encouraging the adoption of sustainable soilless farming practices [5, 9, 139]. In addition, various real-world projects and plans have been implemented globally to leverage soilless technologies. From urban farming initiatives in densely populated cities to large-scale agricultural projects in arid regions, soilless culture systems are used to address food security challenges while promoting environmental sustainability. The following **Table 5** highlights some notable examples of these projects. In addition, for further information related to some websites and applications for hydroponic growers, increase knowledge and simplify soilless cultivation, as mentioned in **Table A1**.

10. Conclusion

In conclusion, as the global population continues to grow and climate change threatens traditional agricultural systems, this chapter introduces the soilless culture techniques as a promising path toward more sustainable food production. However, realizing their full potential requires a holistic approach that addresses environmental, economic, and social considerations. By continually improving the efficiency and sustainability of these systems, we can work toward a future where soilless agriculture plays a crucial role in achieving global food security and environmental sustainability, besides overcoming the major issues of the world, such as climate change, water scarcity, desertification, and non-arable lands. In addition, this chapter highlighted the application of non-conventional sources of nutrients instead of water, besides the alignment between these systems and the Sustainable Development Goals (SDGs), emphasizing their potential to contribute significantly to global efforts in achieving a more sustainable, equitable, and resilient future.

Acknowledgements

Tarek Mahrous is funded by a scholarship [2023GBJ000529] under the joint (Executive program between the Arab Republic of Egypt and China).

Ali Abdelkader is funded by a Chinese government scholarship (CSC) No. [2019GBJ003268].

Conflict of interest

The authors declare no conflict of interest.

DOI: http://dx.doi.org/10.5772/intechopen.1010669

A. Appendix

Tool name	Type	Information	Link	webpage
Hydro Buddy v1.100	website	It is a completely free and open-source program for the calculation of nutrient solutions for hydroponics and general agriculture	https://scienceinhydroponics.com/2016/03/the-first-free-hydroponic-nutrient-calculator-program-o.html	Science in hydroponics
Plantix	App.	It is a free application on smartphones for crop diagnosis and treatment	https://plantix.net/en/	Plantix
USDA National Agricultural Library	website	An official website of the United States government introduces valuable information, videos, and installation steps related to soilless techniques, especially hydroponics	https://www.nal.usda.gov/farms-and-agricultural-production-systems/hydroponics	Hydroponics-National agricultural library
Let Pot	App.	Application on smartphones used for IoT smart controlling in many parameters, such as adjusting brightness, light schedules, and irrigation settings, and the schedule includes a free Plant Diary for tracking growth and a comprehensive Encyclopedia for plant care knowledge to help you grow better and easier	https://letpot.com/?srsltid=AfmBOopGRXJ_YUTFSzVHB8mQkuATj8BE3dpEqjYAD1qSqLZIMc4TO3rt	Let Pot
BudLabs	App.	An application for nutrient calculations for growing to create and update the plants' feeding schedule. It can create hundreds of different feeding chart recipes designed to maximize peak performance and the yield of plants	https://www.advancednutrients.com/nutrient-calculator/	advanced nutrients

Table A1.
Best websites and applications for hydroponic growers, boost knowledge, and simplify soilless cultivation.

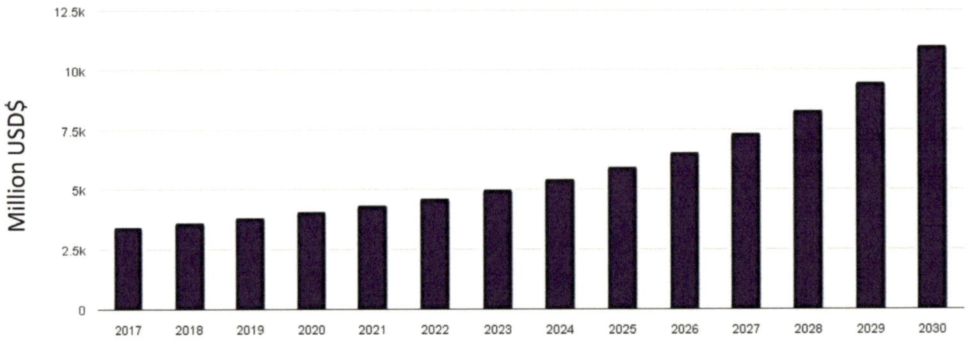

Figure A1.
Global hydroponics market size between years 2017 and 2030 (US$M).

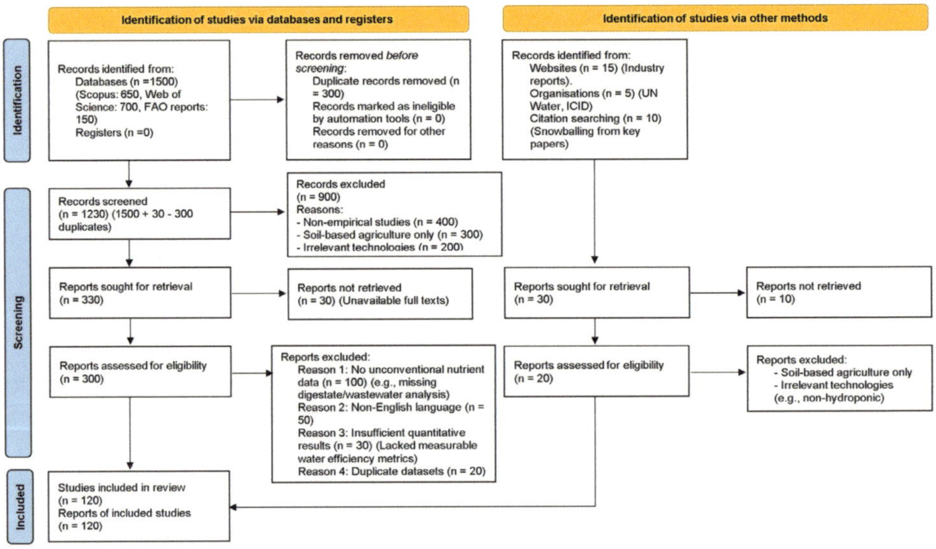

Figure A2.
PRISMA flow diagram for the chapter literature included searches of databases and other sources, indicating the study selection stages and analyzed results.

DOI: http://dx.doi.org/10.5772/intechopen.1010669

Author details

Tarek Mahrous[1,2*] and Ali Mohammed Abd-Elkader[3,4*]

1 Agricultural Engineering Department, Faculty of Agriculture, Cairo University, Giza, Egypt

2 College of Engineering, Huazhong Agricultural University, Wuhan, China

3 Agricultural Botany Department, Faculty of Agriculture, Ain Shams University, Cairo, Egypt

4 Microelement Research Center, College of Resources and Environment, Huazhong Agricultural University, Wuhan, Hubei, China

*Address all correspondence to: tarek.mah.korany@agr.cu.edu.eg
and aliabdelkader@agr.asu.edu.eg

IntechOpen

© 2025 The Author(s). Licensee IntechOpen. This chapter is distributed under the terms of the Creative Commons Attribution License (http://creativecommons.org/licenses/by/4.0), which permits unrestricted use, distribution, and reproduction in any medium, provided the original work is properly cited.

References

[1] Biswas A, Sarkar S, Das S, Dutta S, Roy Choudhury M, Giri A, et al. Water scarcity: A global hindrance to sustainable development and agricultural production – A critical review of the impacts and adaptation strategies. Cambridge Prisms: Water. 2025;**3**:e4. DOI: 10.1017/wat.2024.16

[2] Li L. Water Scarcity, the Climate Crisis and Global Food Security: A Call for Collaborative Action. UN Chronicle Page; 2023. Available from: https://www.un.org/en/un-chronicle/water-scarcity-climate-crisis-and-global-food-security-call-collaborative-action

[3] FAO. The State of World Fisheries and Aquaculture 2022. Rome, Italy: Food and Agriculture Organization; 2022. DOI: 10.4060/cc0461en

[4] Froehlich HE, Gentry RR, Halpern BS. Global change in marine aquaculture production potential under climate change. Nature Ecology and Evolution. 2018;**2**(11):1745-1750. DOI: 10.1038/s41559-018-0669-1

[5] Resh HM. Hydroponic Food Production: A Definitive Guidebook for the Advanced Home Gardener and the Commercial Hydroponic Grower. Boca Raton, Florida, USA: CRC Press; 2022. DOI: 10.1201/9781003133254

[6] Schmautz Z, Walser J-C, Espinal CA, Gartmann F, Scott B, Pothier JF, et al. Microbial diversity across compartments in an aquaponic system and its connection to the nitrogen cycle. Science of the Total Environment. 2022;**852**:158426. DOI: 10.1016/j.scitotenv.2022.158426

[7] Yogev U, Vogler M, Nir O, Londong J, Gross A. Phosphorous recovery from a novel recirculating aquaculture system followed by its sustainable reuse as a fertilizer. Science of the Total Environment. 2020;**722**:137949. DOI: 10.1016/j.scitotenv.2020.137949

[8] Fuentes-Peñailillo F, Gutter K, Vega R, Silva GC. New generation sustainable technologies for soilless vegetable production. Horticulturae. 2024;**10**(1):49. DOI: 10.20944/preprints202311.0488.v1

[9] Mir YH, Mir S, Ganie MA, Shah AM, Majeed U, Chesti MH, et al. Soilless farming: An innovative sustainable approach in agriculture. Pharma Innovation Journal. 2022;**11**(6):2663-2675. Available from: https://www.thepharmajournal.com/special-issue?year=2022&vol=11&issue=6S&ArticleId=13504

[10] Folorunso EA, Schmautz Z, Gebauer R, Mraz J. The economic viability of commercial-scale hydroponics: Nigeria as a case study. Heliyon. 2023;**9**(8):e18979. DOI: 10.1016/j.heliyon.2023.e18979

[11] Goddek S, Joyce A, Kotzen B, Burnell GM. Aquaponics Food Production Systems: Combined Aquaculture and Hydroponic Production Technologies for the Future. Cham, Switzerland: Springer Nature; 2019. DOI: 10.1007/978-3-030-15943-6

[12] Groenveld T, Kohn YY, Gross A, Lazarovitch N. Optimization of nitrogen use efficiency by means of fertigation management in an integrated aquaculture-agriculture system. Journal of cleaner production. 2019;**212**:401-408. DOI: 10.1016/j.jclepro.2018.12.031

[13] FAO, IFAD, UNICEF, WFP and WHO. The State of Food Security and Nutrition

in the World 2021. Transforming Food Systems for Food Security, Improved Nutrition and Affordable Healthy Diets for All. Rome: FAO; 2021. 240 p. DOI: 10.4060/cb4474en

[14] Grand View Research. Global Hydroponics Market Size & Outlook, 2023-2030. 2024. Available from: https://www.grandviewresearch.com/horizon/outlook/hydroponics-market-size/global

[15] Magwaza ST, Magwaza LS, Odindo AO, Mditshwa A, Buckley C. Partially treated domestic wastewater as a nutrient source for tomatoes (Lycopersicum solanum) grown in a hydroponic system: Effect on nutrient absorption and yield. Heliyon. 2020;**6**(12). DOI: 10.1016/j.heliyon.2020.e05745

[16] Rockström J, Edenhofer O, Gaertner J, DeClerck F. Planet-proofing the global food system. Nature Food. 2020;**1**(1):3-5. DOI: 10.1038/s43016-019-0010-4

[17] Kozai T, Niu G, Takagaki M. Plant Factory: An Indoor Vertical Farming System for Efficient Quality Food Production. London, United Kingdom: Academic Press; 2019. DOI: 10.1016/C2014-0-01039-8

[18] Orsini F, Dubbeling M, De Zeeuw H, Gianquinto G. Rooftop Urban Agriculture. Cham, Switzerland: Springer; 2017. DOI: 10.1007/978-3-319-57720-3

[19] Chaudhary A, Anand S. Soilless cultivation: A distinct vision for sustainable agriculture. In: Pandey K, Kushwaha NL, Pande CB, Singh KG, editors. Artificial Intelligence and Smart Agriculture: Technology and Applications. Singapore: Springer Nature Singapore; 2024. pp. 337-368. DOI: 10.1007/978-981-97-0341-8_17

[20] Hussain A, Iqbal K, Aziem S, Mahato P, Negi AK. A review on the science of growing crops without soil (soilless culture)-a novel alternative for growing crops. International Journal of Agriculture and Crop Sciences. 2014;**7**(11):833

[21] Olympios CM, Choukr-Allah R. Overview of soilless culture: Advantages, constraints, and perspectives. Protected Cultivation in the Mediterranean Region. 1999;**31**:307-324

[22] Son JE, Kim HJ, Ahn TI. Chapter 20 - Hydroponic systems. In: Kozai T, Niu G, Takagaki M, editors. Plant Factory. 2nd ed. London: Academic Press; 2020. pp. 273-283. DOI: 10.1016/B978-0-12-816691-8.00020-0

[23] Khan FA. A review on hydroponic greenhouse cultivation for sustainable agriculture. International Journal of Agriculture Environment and Food Sciences. 2018;**2**(2):59-66. DOI: 10.31015/jaefs.18010

[24] Kozai T, Niu G, Masabni JG. Plant Factory Basics, Applications and Advances. London: Academic Press; 2021. Available from: https://www.sciencedirect.com/book/9780323851527

[25] George P, George N. Hydroponics-(soilless cultivation of plants) for biodiversity conservation. International Journal of Modern Trends in Engineering and Research. 2016;**3**(06):97-104

[26] Hosseinzadeh S, Verheust Y, Bonarrigo G, Van Hulle S. Closed hydroponic systems: Operational parameters, root exudates occurrence and related water treatment. Reviews in Environmental Science and Bio/Technology. 2017;**16**(1):59-79. DOI: 10.1007/s11157-016-9418-6

[27] Rajendran S, Domalachenpa T, Arora H, Li P, Sharma A, Rajauria G. Hydroponics: Exploring innovative sustainable technologies and applications across crop production, with emphasis on potato mini-tuber cultivation. Heliyon. 2024;**10**(5):e26823. DOI: 10.1016/j.heliyon.2024.e26823

[28] Mohapatra BC, Chandan NK, Panda SK, Majhi D, Pillai BR. Design and development of a portable and streamlined nutrient film technique (NFT) aquaponic system. Aquacultural Engineering. 2020;**90**:102100. DOI: 10.1016/j.aquaeng.2020.102100

[29] Ahamed MS, Sultan M, Shamshiri RR, Rahman MM, Aleem M, Balasundram SK. Present status and challenges of fodder production in controlled environments: A review. Smart Agricultural Technology. 2023;**3**:100080. DOI: 10.1016/j.atech.2022.100080

[30] Frasetya B, Harisman K, Ramdaniah NAH. The Effect of Hydroponics Systems on the Growth of lettuce2021. Bandung Barat, Indonesia: IOP Publishing; 2021. DOI: 10.1088/1757-899X/1098/4/042115

[31] Ferrarezi RS, Testezlaf R. Performance of wick irrigation system using self-compensating troughs with substrates for lettuce production. Journal of Plant Nutrition. 2016;**39**(1):147-161. DOI: 10.1080/01904167.2014.983127

[32] Mohamed TMK, Gao J, Abuarab ME, Kassem M, Wasef E, El-Ssawy W. Applying different magnetic water densities as irrigation for aeroponically and hydroponically grown strawberries. Agriculture. 2022;**12**(6):819. DOI: 10.3390/agriculture12060819

[33] Mohamed TMK, Gao J, Tunio M. Development and experiment of the intelligent control system for rhizosphere temperature of aeroponic lettuce via the internet of things. International Journal of Agricultural and Biological Engineering. 2022;**15**(3):225-233. DOI: 10.25165/j.ijabe.20221503.6530

[34] Yep B, Zheng Y. Aquaponic trends and challenges–a review. Journal of Cleaner Production. 2019;**228**:1586-1599. DOI: 10.1016/j.jclepro.2019.04.290

[35] Sardare MD, Admane SV. A review on plant without soil-hydroponics. International Journal of Research in Engineering and Technology. 2013;**2**(3):299-304. DOI: 10.15623/ijret.2013.0203013

[36] Colt J, Schuur AM, Weaver D, Semmens K. Engineering design of aquaponics systems. Reviews in Fisheries Science and Aquaculture. 2022;**30**(1):33-80. DOI: 10.1080/23308249.2021.1886240

[37] Yanes AR, Martinez P, Ahmad R. Towards automated aquaponics: A review on monitoring, IoT, and smart systems. Journal of Cleaner Production. 2020;**263**:121571. DOI: 10.1016/j.jclepro.2020.121571

[38] Love DC, Fry JP, Li X, Hill ES, Genello L, Semmens K, et al. Commercial aquaponics production and profitability: Findings from an international survey. Aquaculture. 2015;**435**:67-74. DOI: 10.1016/j.aquaculture.2014.09.023

[39] Rajaseger G, Chan KL, Yee Tan K, Ramasamy S, Khin MC, Amaladoss A, et al. Hydroponics: Current trends in sustainable crop production. Bioinformation. 2023;**19**(9):925-938. DOI: 10.6026/97320630019925

[40] Dbritto G, Hamdare S, editors. An AI Based System Design to Develop and Monitor a Hydroponic Farm. 2018 International Conference on Smart City and Emerging Technology (ICSCET); 5-5 January 2018; 2018. DOI: 10.1109/ICSCET.2018.8537317

[41] Pomoni DI, Koukou MK, Vrachopoulos MG, Vasiliadis L. A review of hydroponics and conventional agriculture based on energy and water consumption, environmental impact, and land use. Energies. 2023;**16**(4):1690. DOI: 10.3390/en16041690

[42] Adão T, Hruška J, Pádua L, Bessa J, Peres E, Morais R, et al. Hyperspectral imaging: A review on UAV-based sensors, data processing and applications for agriculture and forestry. Remote Sensing. 2017;**9**(11):1110. DOI: 10.3390/rs9111110

[43] Sadek N, kamal N, Shehata D. Internet of things based smart automated indoor hydroponics and aeroponics greenhouse in Egypt. Ain Shams Engineering Journal. 2024;**15**(2):102341. DOI: 10.1016/j.asej.2023.102341

[44] Ghosh B, Roy S, Ahmed N, De D. Dew Aeroponics: Dew-Enabled Smart Aeroponics System in Agriculture 4.0. Dew Computing: The Sustainable IoT Perspectives. Singapore: Springer; 2023. pp. 261-287. DOI: 10.1007/978-981-99-4590-0_13

[45] Rahman MA, Chakraborty NR, Sufiun A, Banshal SK, Tajnin FR. An AIoT-based hydroponic system for crop recommendation and nutrient parameter monitorization. Smart Agricultural Technology. 2024;**8**:100472. DOI: 10.1016/j.atech.2024.100472

[46] Pitakphongmetha J, Boonnam N, Wongkoon S, Horanont T, Somkiadcharoen D, Prapakornpilai J, editors. Internet of Things for Planting in Smart Farm Hydroponics Style. 2016 International Computer Science and Engineering Conference (ICSEC); 14-17 December 2016; 2016. DOI: 10.1109/ICSEC.2016.7859872

[47] Garzón J, Montes L, Garzón J, Lampropoulos G. Systematic review of technology in aeroponics: Introducing the technology adoption and integration in sustainable agriculture model. Agronomy. 2023;**13**(10):2517. DOI: 10.3390/agronomy13102517

[48] Idoje G, Mouroutoglou C, Dagiuklas T, Kotsiras A, Muddesar I, Alefragkis P. Comparative analysis of data using machine learning algorithms: A hydroponics system use case. Smart Agricultural Technology. 2023;**4**:100207. DOI: 10.1016/j.atech.2023.100207

[49] Mamatha V, Kavitha JC. Machine learning based crop growth management in greenhouse environment using hydroponics farming techniques. Measurement: Sensors. 2023;**25**:100665. DOI: 10.1016/j.measen.2023.100665

[50] Maraveas C, Bartzanas T. Application of internet of things (IoT) for optimized greenhouse environments. AgriEngineering. 2021;**3**(4):954-970. DOI: 10.1016/j.compag.2022.106993

[51] Naresh R, Jadav SK, Singh M, Patel A, Singh B, Beese S, et al. Role of hydroponics in improving water-use efficiency and food security. International Journal of Environment and Climate Change. 2024;**14**(2):608-633. DOI: 10.9734/ijecc/2024/v14i23976c

[52] Amprako JL. The United Nations World Water Development Report 2015: Water for a Sustainable World.

2016. pp. 64-65. Available from: https://digitallibrary.un.org/record/3892926

[53] Pereira LS, Paredes P, Hunsaker DJ, López-Urrea R, Jovanovic N. Updates and advances to the FAO56 crop water requirements method. Agricultural Water Management. 2021;**248**:106697. DOI: 10.1016/J.AGWAT.2020.106697

[54] Chapagain AK, Hoekstra AY. The Water Needed to Have the Dutch Drink Coffee. The Netherlands: Unesco-IHE; 2003. Available from: https://www.waterfootprint.org/resources/Report14.pdf

[55] Ashine ET, Bedane MT. A review on factors affecting the uniformity of sprinkler irrigation. Engineering and Applied Sciences. 2022;**7**(5):71-76. DOI: 10.11648/j.eas.20220705.13

[56] Jägermeyr J, Gerten D, Heinke J, Schaphoff S, Kummu M, Lucht W. Water savings potentials of irrigation systems: Global simulation of processes and linkages. Hydrology and Earth System Sciences. 2015;**19**(7):3073-3091. DOI: 10.5194/hess-19-3073-2015

[57] Jalota SK, Prihar SS. Reducing Soil Water Evaporation with Tillage and Straw Mulching. London: Eurospan Group; 1998. Available from: https://www.cabidigitallibrary.org/doi/full/10.5555/19981906474

[58] Goddek S, Delaide B, Mankasingh U, Ragnarsdottir KV, Jijakli H, Thorarinsdottir R. Challenges of sustainable and commercial aquaponics. Sustainability. 2015;**7**(4):4199-4224. DOI: 10.3390/su7044199

[59] Jensen MH. Controlled Environment Agriculture in Deserts, Tropics and Temperate Regions - A World review. Leuven, Belgium: International Society for Horticultural Science (ISHS); 2002. DOI: 10.17660/ActaHortic.2002.578.1

[60] O'Connor J, Mickan BS, Rinklebe J, Song H, Siddique KHM, Wang H, et al. Environmental implications, potential value, and future of food-waste anaerobic digestate management: A review. Journal of Environmental Management. 2022;**318**:115519. DOI: 10.1016/j.jenvman.2022.115519

[61] Kozai T. Resource use efficiency of closed plant production system with artificial light: Concept, estimation and application to plant factory. Proceedings of the Japan Academy, Series B. 2013;**89**(10):447-461. DOI: 10.2183/pjab.89.447

[62] Nwanojuo MA, Anumudu CK, Onyeaka H. Impact of controlled environment agriculture (CEA) in Nigeria, a review of the future of farming in Africa. Agriculture. 2025;**15**(2):117. DOI: 10.3390/agriculture15020117

[63] Gómez C, Currey CJ, Dickson RW, Kim H-J, Hernández R, Sabeh NC, et al. Controlled environment food production for urban agriculture. HortScience. 2019;**54**(9):1448-1458. DOI: 10.21273/HORTSCI14073-19

[64] Benke K, Tomkins B. Future food-production systems: Vertical farming and controlled-environment agriculture. Sustainability: Science, Practice and Policy. 2017;**13**(1):13-26. DOI: 10.1080/15487733.2017.1394054

[65] Mitchell CA, Dzakovich MP, Gomez C, Lopez R, Burr JF, Hernández R, et al. Light-emitting diodes in horticulture. Horticultural Reviews. 2015;**43**:1-88. DOI: 10.1002/9781119107781.ch01

[66] Graamans L, Baeza E, Van Den Dobbelsteen A, Tsafaras I, Stanghellini C. Plant factories versus greenhouses: Comparison of resource use efficiency. Agricultural Systems. 2018;**160**:31-43. DOI: 10.1016/j.agsy.2017.11.003

[67] van Delden SH, SharathKumar M, Butturini M, Graamans LJA, Heuvelink E, Kacira M, et al. Current status and future challenges in implementing and upscaling vertical farming systems. Nature Food. 2021;**2**(12):944-956. DOI: 10.1038/s43016-021-00402-w

[68] Li D, Li C, Yao Y, Li M, Liu L. Modern imaging techniques in plant nutrition analysis: A review. Computers and Electronics in Agriculture. 2020;**174**:105459. DOI: 10.1016/j.compag.2020.105459

[69] Boyko H. Salinity and aridity new approaches to old problems. In: Boyko H, editor. Salinity and Aridity: New Approaches to Old Problems. Dordrecht: Springer Netherlands; 1966. pp. 3-22. DOI: 10.1007/978-94-017-6014-0_1

[70] Hoagland DR, Arnon DI. The Water-Culture Method for Growing Plants without Soil. California: California Agricultural Experiment Station Circular-347; 1938

[71] Atzori G, Mancuso S, Masi E. Seawater potential use in soilless culture: A review. Scientia Horticulturae. 2019;**249**:199-207. DOI: 10.1016/j.scienta.2019.01.035

[72] Glenn EP, Brown JJ, Blumwald E. Salt tolerance and crop potential of halophytes. Critical Reviews in Plant Sciences. 1999;**18**(2):227-255. DOI: 10.1080/07352689991309207

[73] Jankauskienė J, Laužikė K, Kaupaitė S. The use of anaerobic digestate for greenhouse horticulture. Agronomy. 2024;**14**(10):2437. DOI: 10.3390/agronomy14102437

[74] Kovačić Đ, Lončarić Z, Jović J, Samac D, Popović B, Tišma M. Digestate management and processing practices: A review. Applied Sciences. 2022;**12**(18):9216. DOI: 10.3390/app12189216

[75] Larsen TA, Maurer M, Udert KM, Lienert J. Nutrient cycles and resource management: Implications for the choice of wastewater treatment technology. Water Science and Technology. 2007;**56**(5):229-237. DOI: 10.2166/wst.2007.576

[76] Maurer M, Schwegler P, Larsen TA. Nutrients in urine: Energetic aspects of removal and recovery. Water Science and Technology. 2003;**48**(1):37-46. DOI: 10.2166/wst.2003.0011

[77] Breckle S-W. Is sustainable agriculture with seawater irrigation realistic? In: Ashraf M, Ozturk M, Athar HR, editors. Salinity and Water Stress: Improving Crop Efficiency. Dordrecht: Springer Netherlands; 2009. pp. 187-196. DOI: 10.1007/978-1-4020-9065-3_19

[78] de Vos A, Bruning B, van Straten G, Oosterbaan R, Rozema J, van Bodegom P. Crop Salt Tolerance under Controlled Field Conditions in the Netherlands, Based on Trials Conducted at Salt Farm Texel. Geneva, The Netherlands: Salt Farm Texel; 2016. Available from: https://edepot.wur.nl/409817

[79] Ventura Y, Eshel A, Pasternak D, Sagi M. The development of halophyte-based agriculture: Past and present. Annals of Botany. 2015;**115**(3):529-540. DOI: 10.1093/aob/mcu173

[80] Pardossi A, Tognoni F, Incrocci L. Mediterranean greenhouse technology.

Chronica Horticulturae. 2004;**44**(2):28-34. Available from: https://ceac.arizona.edu/sites/default/files/Mediterranean.pdf

[81] Lamolinara B, Pérez-Martínez A, Guardado-Yordi E, Guillén Fiallos C, Diéguez-Santana K, Ruiz-Mercado GJ. Anaerobic digestate management, environmental impacts, and techno-economic challenges. Waste Management. 2022;**140**:14-30. DOI: 10.1016/j.wasman.2021.12.035

[82] Monlau F, Sambusiti C, Ficara E, Aboulkas A, Barakat A, Carrère H. New opportunities for agricultural digestate valorization: Current situation and perspectives. Energy and Environmental Science. 2015;**8**(9):2600-2621. DOI: 10.1039/c5ee01633a

[83] Pelayo Lind O, Hultberg M, Bergstrand K-J, Larsson-Jönsson H, Caspersen S, Asp H. Biogas digestate in vegetable hydroponic production: pH dynamics and pH management by controlled nitrification. Waste and Biomass Valorization. 2021;**12**:123-133. DOI: 10.1007/s12649-020-00965-y

[84] Rubert A, Costa JA, Colla LM, Hemkemeier M. Valorization of liquid digestate from wastewater and microalgae: A promising proposal for nutrient recovery in hydroponic systems. Environment, Development and Sustainability. 2024:1-36. DOI: 10.1007/s10668-024-04726-y

[85] Vaneeckhaute C, Lebuf V, Michels E, Belia E, Vanrolleghem PA, Tack FMG, et al. Nutrient recovery from digestate: Systematic technology review and product classification. Waste and Biomass Valorization. 2017;**8**:21-40. DOI: 10.1007/s12649-016-9642-x

[86] Möller K, Müller T. Effects of anaerobic digestion on digestate nutrient availability and crop growth: A review. Engineering in Life Sciences. 2012;**12**(3):242-257. DOI: 10.1002/elsc.201100085

[87] Lai C, Tu M, Li M, Yu S. Remarkable solvent and extractable lignin effects on enzymatic digestibility of organosolv pretreated hardwood. Bioresource Technology. 2014;**156**:92-99. DOI: 10.1016/j.biortech.2014.01.030

[88] Malik SJ, Kasulla S, Zafar S, Bhatta AD, Elboughdiri N. Using Biogas Plant Slurry to Optimize pH Dynamics and Control Nitrification in Hydroponic Systems. Available at SSRN 49675002024. DOI: 10.5281/zenodo.13827526

[89] Weimers K, Bergstrand K-J, Hultberg M, Asp H. Liquid anaerobic digestate as sole nutrient source in soilless horticulture—Or spiked with mineral nutrients for improved plant growth. Frontiers in Plant Science. 2022;**13**:770179. DOI: 10.3389/fpls.2022.770179

[90] Wang L, Guo S, Wang Y, Yi D, Wang J. Poultry biogas slurry can partially substitute for mineral fertilizers in hydroponic lettuce production. Environmental Science and Pollution Research. 2019;**26**:659-671. Available from: https://link.springer.com/article/10.1007/s11356-018-3538-1

[91] Yang L, Giannis A, Chang VWC, Liu B, Zhang J, Wang J-Y. Application of hydroponic systems for the treatment of source-separated human urine. Ecological Engineering. 2015;**81**:182-191. DOI: 10.1016/j.ecoleng.2015.04.013

[92] Bliedung A, Dockhorn T, Germer J, Mayerl C, Mohr M. Experiences of

running a hydroponic system in a pilot scale for resource-efficient water reuse. Journal of Water Reuse and Desalination. 2020;**10**(4):347-362. DOI: 10.2166/wrd.2020.014

[93] Volpin F, Yu H, Cho J, Lee C, Phuntsho S, Ghaffour N, et al. Human urine as a forward osmosis draw solution for the application of microalgae dewatering. Journal of Hazardous Materials. 2019;**378**:120724. DOI: 10.3390/agriculture12050646

[94] Jurga A, Janiak K, Wizimirska A, Chochura P, Miodoński S, Muszyński-Huhajło M, et al. Resource recovery from synthetic nitrified urine in the hydroponic cultivation of lettuce (Lactuca sativa Var. capitata L.). Agronomy. 2021;**11**(11):2242. DOI: 10.3390/agronomy11112242

[95] Cifuentes-Torres L, Mendoza-Espinosa LG, Correa-Reyes G, Daesslé LW. Hydroponics with wastewater: A review of trends and opportunities. Water and Environment Journal. 2021;**35**(1):166-180. DOI: 10.1111/wej.12617

[96] Rana S, Bag SK, Golder D, Roy SM, Pradhan C, Jana BB. Reclamation of municipal domestic wastewater by aquaponics of tomato plants. Ecological Engineering. 2011;**37**(6):981-988. DOI: 10.1016/j.ecoleng.2011.01.009

[97] Al-Karaki GN. Utilization of treated sewage wastewater for green forage production in a hydroponic system. Emirates Journal of Food and Agriculture (EJFA). 2011;**23**(1):80-94. DOI: 10.9755/EJFA.V23I1.5315

[98] Júnior JAS, Gheyi HR, da Silva DN, Araujo DL, Guedes Filho DH. Substratos e diferentes concentrações da solução nutritiva preparada em água residuária no crescimento do girassol. Revista Ciência Agronômica. 2014;**45**(4):696-707. DOI: 10.1590/S1806-66902014000400007

[99] Lopez-Galvez F, Gil MI, Pedrero-Salcedo F, Alarcón JJ, Allende A. Monitoring generic Escherichia coli in reclaimed and surface water used in hydroponically cultivated greenhouse peppers and the influence of fertilizer solutions. Food Control. 2016;**67**:90-95. DOI: 10.1016/j.foodcont.2016.02.037

[100] Sundar P, Jyothi K, Sundar C. Indoor hydroponics: A potential solution to reuse domestic rinse water. Biosciences Biotechnology Research Asia. 2021;**18**(2):373-383. DOI: 10.13005/bbra/2924

[101] Al Hamedi F, Karthishwaran K, Alyafei MAM. Hydroponic wheat production using fresh water and treated wastewater under the semi-arid region. Emirates Journal of Food and Agriculture. 2021;**33**:178. DOI: 10.9755/ejfa.2021.v33.i2.2620

[102] Carvalho RSC, Bastos RG, Souza CF. Influence of the use of wastewater on nutrient absorption and production of lettuce grown in a hydroponic system. Agricultural Water Management. 2018;**203**:311-321. DOI: 10.1016/j.agwat.2018.03.028

[103] Cuba RS, do Carmo JR, Souza CF, Bastos RG. Potencial de efluente de esgoto doméstico tratado como fonte de água e nutrientes no cultivo hidropônico de alface. Revista Ambiente and Água. 2015;**10**(3):574-586. DOI: 10.4136/ambi-agua.1575

[104] Damasceno LMO, Andrade Júnior AS, Gheyi HR. Cultivation of gerbera irrigated with treated domestic effluents. Revista Brasileira de Engenharia Agrícola e Ambiental.

2010;**14**:582-588. DOI: 10.1590/S1415-43662010000600003

[105] Ceci L, Cavalera MA, Serrapica F, Di Francia A, Masucci F, Carelli G. Use of reclaimed urban wastewater for the production of hydroponic barley forage: Water characteristics, feed quality and effects on health status and production of lactating cows. Frontiers in Veterinary Science. 2023;**10**:1274466. DOI: 10.3389/fvets.2023.1274466

[106] Krishnasamy K, Nair J, Bäuml B. Hydroponic system for the treatment of anaerobic liquid. Water Science and Technology. 2012;**65**(7):1164-1171. DOI: 10.2166/wst.2012.031

[107] Haddad M, Mizyed N. Evaluation of various hydroponic techniques as decentralised wastewater treatment and reuse systems. International Journal of Environmental Studies. 2011;**68**(4):461-476. DOI: 10.1080/00207233.2011.582701

[108] Keeratiurai P. Efficiency of wastewater treatment with hydroponics. ARPN Journal of Agricultural and Biological Science. 2013;**8**(12):800-805

[109] Magwaza ST, Magwaza LS, Odindo AO, Mditshwa A. Hydroponic technology as decentralised system for domestic wastewater treatment and vegetable production in urban agriculture: A review. Science of the Total Environment. 2020;**698**:134154. DOI: 10.1016/j.scitotenv.2019.134154

[110] Rababah AA, Ashbolt NJ. Innovative production treatment hydroponic farm for primary municipal sewage utilisation. Water Research. 2000;**34**(3):825-834. DOI: 10.1016/S0043-1354(99)00231-6

[111] Vaillant N, Monnet F, Sallanon H, Coudret A, Hitmi A. Use of commercial plant species in a hydroponic system to treat domestic wastewaters. Journal of environmental quality. 2004;**33**(2):695-702. DOI: 10.2134/jeq2004.6950

[112] Yang Z, Zheng S, Chen J, Sun M. Purification of nitrate-rich agricultural runoff by a hydroponic system. Bioresource Technology. 2008;**99**(17):8049-8053. DOI: 10.1016/j.biortech.2008.03.040

[113] Alcalde-Sanz L, Gawlik BM. Minimum quality requirements for water reuse in agricultural irrigation and aquifer recharge. In: Towards a Water Reuse Regulatory Instrument at EU Level. Publications Office of the European Union; 2017. p. 57. Available from: https://data.europa.eu/doi/10.2760/887727

[114] World Health Organization. World Health Organization Guidelines for the Safe Use of Wasterwater Excreta and Greywater. Geneva, Switzerland: World Health Organization; 2006. Available from: https://www.who.int/publications/i/item/9241546859

[115] Singh A, Ashoka P, Tyagi V, Panotra N, Kalpana K, Pandey SK. A comparative analysis of land, water and energy requirements for hydroponic and conventional cultivation of horticultural crops. Journal of Scientific Research and Reports. 2024;**30**(9):771-784. DOI: 10.9734/jsrr/2024/v30i92404

[116] Group L. How Much Electricity Does a Hydroponic System Use. 2024. Available from: https://www.lyinegroup.com/a-how-much-electricity-does-a-hydroponic-system-use.html

[117] Fathidarehnijeh E, Nadeem M, Cheema M, Thomas R, Krishnapillai M, Galagedara L. Current perspective on nutrient solution management strategies to improve the nutrient and water use efficiency in hydroponic systems. Canadian Journal of Plant Science. 2023;**104**(2):88-102. DOI: 10.1139/cjps-2023-0034

[118] Regmi A, Rueda-Kunz D, Liu H, Trevino J, Kathi S, Simpson C. Comparing resource use efficiencies in hydroponic and aeroponic production systems. Technology in Horticulture. 2024;**4**(1):10.48130/tihort-0024-0002

[119] Dutta M, Gupta D, Sahu S, Limkar S, Singh P, Mishra A, et al. Evaluation of growth responses of lettuce and energy efficiency of the substrate and smart hydroponics cropping system. Sensors (Basel, Switzerland). 2023;**23**(4):1875. DOI: 10.3390/s23041875

[120] Wong AYK, Ya'akub AR, JAM G, Teh AA, Kovtun V, Pusppanathan J, et al. Greening of automated hydroponic system based on IoT and enhanced solar energy with electricity saving box. Journal of Advanced Research in Applied Sciences and Engineering Technology. 2024;**62**:142-150. DOI: 10.37934/araset.62.1.142150

[121] Caputo S. History, Techniques and Technologies of Soil-Less Cultivation. Small Scale Soil-Less Urban Agriculture in Europe. Cham: Springer International Publishing; 2022. pp. 45-86. DOI: 10.1007/978-3-030-99962-9_4

[122] Mishra SJ, Rout D, Sahoo D. Analysing the economic viability of hydroponic farming: A comparative cost-benefit analysis. International Journal of Progressive Research in Engineering Management and Science (IJPREMS). 2024;**4**(6):1806-1811. DOI: 10.58257/IJPREMS35128

[123] Savvas D, Gruda N. Application of soilless culture technologies in the modern greenhouse industry—A review. European Journal of Horticultural Science. 2018;**83**(5):280-293. DOI: 10.17660/eJHS.2018/83.5.2

[124] Tüzel Y, Gül A, Tüzel IH, Öztekin GB. Different soilless culture systems and their management. Journal of Agricultural, Food and Environmental Sciences, JAFES. 2019;**73**(3):7-12. DOI: 10.55302/JAFES19733007t

[125] Wang W, Chang J-S, Lee D-J. Anaerobic digestate valorization beyond agricultural application: Current status and prospects. Bioresource Technology. 2023;**373**:128742. DOI: 10.1016/j.biortech.2023.128742

[126] Fussy A, Papenbrock J. An overview of soil and soilless cultivation techniques—Chances, challenges and the neglected question of sustainability. Plants. 2022;**11**:1153. DOI: 10.3390/plants11091153

[127] Gebreegziher WG. Soilless culture technology to transform vegetable farming, reduce land pressure and degradation in drylands. Cogent Food and Agriculture. 2023;**9**(2):2265106. DOI: 10.1080/23311932.2023.2265106

[128] Kumar TV, Verma R. A comprehensive review on soilless cultivation for sustainable agriculture. Journal of Experimental Agriculture International. 2024;**46**(6):193-207. DOI: 10.9734/JEAI/2024/v46i62470

[129] Singh D, Basu C, Meinhardt-Wollweber M, Roth B. LEDs for energy efficient greenhouse lighting. Renewable and Sustainable Energy Reviews. 2015;**49**:139-147. DOI: 10.1016/j.rser.2015.04.117

[130] Gonnella M, Renna M. The evolution of soilless systems towards ecological sustainability in the perspective of a circular economy. Is it really the opposite of organic agriculture? Agronomy. 2021;**11**(5):950. DOI: 10.3390/agronomy11050950

[131] Jia X, Ma P, Wei C-I, Wang Q. Chitin and chitosan: Pioneering sustainable

substrates for next-generation soilless vertical farming. Trends in Food Science and Technology. 2024;**150**:104599. DOI: 10.1016/j.tifs.2024.104599

[132] Al Meselmani MA. Hydroponics: The Future of Sustainable Farming. Hydroponics: The Future of Sustainable Farming. New York: Springer; 2024. pp. 101-122. DOI: 10.1007/978-1-0716-3993-1

[133] Velazquez-Gonzalez RS, Garcia-Garcia AL, Ventura-Zapata E, Barceinas-Sanchez JDO, Sosa-Savedra JC. A review on hydroponics and the technologies associated for medium-and small-scale operations. Agriculture. 2022;**12**(5):646. DOI: 10.3390/agriculture12050646

[134] Tajudeen AL, Taiwo OS. Soilless farming–a key player in the realisation of "zero hunger" of the sustainable development goals in Nigeria. International Journal of Ecology and Environmental Sciences. 2018;**5**:1-7. Available from:http://article.aascit.org/file/pdf/9040836.pdf

[135] Sisodia GS, Alshamsi R, Sergi BS. Business valuation strategy for new hydroponic farm development–a proposal towards sustainable agriculture development in United Arab Emirates. British Food Journal. 2021;**123**(4):1560-1577. DOI: 10.1108/BFJ-06-2020-0557

[136] Khan MM, Younis A, Akram MT, Ijaz MM, Al-Sadi AM. Feeding the cities: Urban agriculture for food security and sustainability of urban areas. CABI Reviews. 2024;**19**(1):1366. DOI: 10.1079/cabireviews.2024.0053

[137] Nicola S, Pignata G, Ferrante A, Bulgari R, Cocetta G, Ertani A. Water use efficiency in greenhouse systems and its application in horticulture. AgroLife Scientific Journal. 2020;**9**(1):248-262. Available from: https://air.unimi.it/handle/2434/740785?mode=complete

[138] Barrett GE, Alexander PD, Robinson JS, Bragg NC. Achieving environmentally sustainable growing media for soilless plant cultivation systems–a review. Scientia Horticulturae. 2016;**212**:220-234. DOI: 10.1016/j.scienta.2016.09.030

[139] Gruda NS, Machado RMA, van Os EA. Is soilless culture a sustainable form of agriculture? Horticulturae. 2023;**9**:1190. DOI: 10.3390/horticulturae9111190

[140] Jones K. The scalability of urban agriculture: Chicago case studies. Enquiry The ARCC Journal for Architectural Research. 2024;**21**(1):1-15. DOI: 10.17831/enqarcc.v21i1.1195

[141] Roz P. The International Community's Role in the Climate and Environment Space in Jordan. Brighton, UK: The Institute of Development Studies and Partner Organisations; 2024. Available from: https://hdl.handle.net/20.500.12413/14570; https://opendocs.ids.ac.uk/articles/report/The_International_Community_s_Role_in_the_Climate_and_Environment_Space_in_Jordan/26433490

[142] Heuvelink E, Hemming S, Marcelis LFM. Some recent developments in controlled-environment agriculture: On plant physiology, sustainability, and autonomous control. The Journal of Horticultural Science and Biotechnology. 2025;**100**:1-11. DOI: 10.1080/14620316.2024.2440592

[143] Dabakarov S, Borowka W, Ricci A. Greenhouse Hydroponics: An Opportunity for Enhanced Academic Learning and Food Sustainability at Skidmore. Capstone

Submitted to the Department of Environmental Studies and Sciences, Skidmore College ("Greenhouse Hydroponics: An Opportunity for Enhanced Academic-Skidmore"). 2020. Available from: https://www.skidmore.edu/environmental_studies/capstone/projects/documents/7-Hydroponics-2020.pdf

[144] Balani HR. Project Management Plan for the Construction of the First Urban Hydroponics Farm within the Residential Community in the king's Park Area of Belize City, Belize [Master's in Project Management (MPM) Degree]. Belize: Universidad Para La Cooperacion Internacional; 2023. Available from: https://omeka.campusuci2.com/biblioteca/items/show/2094

Chapter 5

Sustainable Soil and Water Management in Arid Climates in the Mediterranean Climate Zone

Ertuğrul Karaş

Abstract

Climate change and global warming pose serious threats to agriculture, water resources, and the environment, especially in the Mediterranean climate zone. This region is facing rising temperatures and decreasing rainfall, while the frequency of extreme weather events, such as droughts, floods, and heavy rainfall, is increasing. The Mediterranean basin holds only 1.2% of the world's renewable water resources, and it is expected that freshwater resources will decrease by 25–50% by 2050. Turkey is one of the most affected countries, with diminishing water sources and significant losses in agricultural land, creating serious challenges for both water supply and rural development. Rising temperatures have negatively impacted traditional farming practices. For example, the planting dates of winter crops like wheat and barley have shifted, resulting in lower yields due to water stress. Moreover, water shortages have led to reduced soil moisture, making it harder for crops to grow during critical periods. In Turkey, much of the water used in agriculture comes from groundwater, which is rapidly depleting, causing a decline in groundwater levels and increased energy costs for extraction. To overcome these challenges, sustainable soil and water management practices must be implemented. Efficient irrigation techniques, water-saving methods, and solutions to enhance agricultural productivity are crucial to addressing these issues.

Keywords: global warming, climate change, agricultural production, dry climates, sustainable soil and water management

1. Introduction

Global warming and drought are significant risks to agriculture and food security [1]. These challenges are seen not only in key climate changes such as temperature, precipitation, humidity, and sunlight but also in the increased frequency of extreme weather events, such as floods and droughts. As of 2025, the global population reached 8.2 billion, with 16% living under water stress [2]. Agriculture, which heavily relies on irrigation, uses 70% of both surface and groundwater, contributing 40% of total agricultural output from just 24% of irrigated land [3]. However, only half of the irrigation water is effectively utilized by plants [4], and the demand for irrigation water is expected to rise by 1.7 times by the mid-twenty-first century, complicating agricultural production [5]. With the global population projected to reach 9.7 billion by 2050, the increasing demand for water in domestic and industrial sectors is likely to reduce water availability for agriculture [6].

The Mediterranean region is highly vulnerable to climate change and human activities, especially regarding water scarcity [7]. Meteorological and hydrological droughts are becoming more frequent, with average temperatures rising by 0.1°C per decade since the 1970s, and precipitation decreasing by 25 mm [8]. Projections show that by 2050, the region could experience a temperature rise of 1.5–2.5°C and a 5–20% reduction in rainfall [9], leading to drier conditions and more arid climates, particularly in countries like Italy, Greece, and Turkey [10]. The Mediterranean holds just 1.2% of the world's renewable freshwater resources, with countries like France, Italy, Greece, and Turkey being major sources [11]. However, freshwater resources in the region could decline by 25–50% by 2050, exacerbating the challenges of meeting both drinking and agricultural water demands, particularly due to growing tourism, urbanization, and population density [12].

Turkey, one of the countries most vulnerable to climate change, faces rising temperatures and decreasing precipitation [13]. The population, expected to grow from 51 million in 1985 to 85.6 million by 2025, has resulted in a 14.2% reduction in agricultural land use due to urbanization and industrialization [14]. Wheat cultivation has dropped by 22%, further affecting water availability for agriculture [15]. These changes have led to regional disparities [16], particularly between eastern and southeastern provinces and the rest of the country [17]. Climate change has led to more frequent droughts, heavy rainfall, and floods in Turkey [18], with water stress and land degradation becoming major issues [19]. Overuse of groundwater has caused sinkholes, particularly in Central Anatolia, where declining water levels in 2006 led to an 80% decrease in river flow, severely affecting agricultural and hydrological systems [20]. Over the past 60 years, 186 out of 240 lakes have dried up, further threatening biodiversity and causing a decline in water levels, surface areas, and oxygen content in remaining lakes [21].

The changing climate has also altered growing seasons, especially for winter crops like wheat and barley. Warmer autumns have delayed planting by 20–30 days, leading to reduced germination and water retention during winter months. This results in shorter growing periods, earlier transitions to the generative phase, and lower yields [22]. Additionally, warmer temperatures have allowed for the diversification of crops. For instance, corn, which was once limited to irrigated areas, is now the second most common crop in Central Anatolia.

This chapter outlines the current situation in Central Anatolia and highlights the necessary measures for sustainable soil and water resource management in arid and semiarid climates.

2. Materials

In this section, comprehensive information will be provided about the study area covering aspects such as soil properties, climate, topographic conditions, land use, vegetation, tillage management, irrigation practices, and soil erosion status. These data will help to reveal the current conditions of the study area. The climate data used in this study were obtained from the General Directorate of Meteorology, ensuring accuracy and reliability. Data on soil properties, topography, land use, vegetation, irrigation methods, and erosion status were compiled from published research reports and scientific articles.

This section aims to provide a holistic understanding of the environmental and agricultural context of the study area, which is necessary to analyze and address issues such as soil erosion and land management, tillage, and irrigation, by integrating these various datasets. While some of the data used in the study were obtained from the finalized reports, a significant portion of the statistical data was also simplified by reprocessing these data to serve the purpose. In order to ensure the integrity of the study, a summary of the current situation is given as main headings in the results section.

2.1 Study area

Turkey is divided into seven geographical regions, and each differing in various characteristics. Central Anatolia, as shown in **Figure 1**, is located in the heart of the country and comprises 13 provinces [23]. The region spans an area of 15.1 million hectares, representing 18% of Turkey's total land area, which is 78 million hectares [24].

2.2 Climate characteristics

Turkey is home to three main climate types: Steppe, Mediterranean, and Black Sea. Four geographical regions of the country fall within the steppe climate zone, namely Central Anatolia, Eastern Anatolia, Southeastern Anatolia, and part of Thrace (**Figure 2**). The steppe climate regions together cover roughly 50% of the total land area of Turkey [24].

The steppe climate in Central Anatolia, classified as cold semiarid, is characterized by temperate continental conditions. In this arid or semiarid region, summers are typically hot and dry, while winters are cold and freezing, with the intensity of the cold increasing as one moves eastward [25]. Winters are dry, often accompanied

Figure 1.
Location of Central Anatolia in Turkey [23].

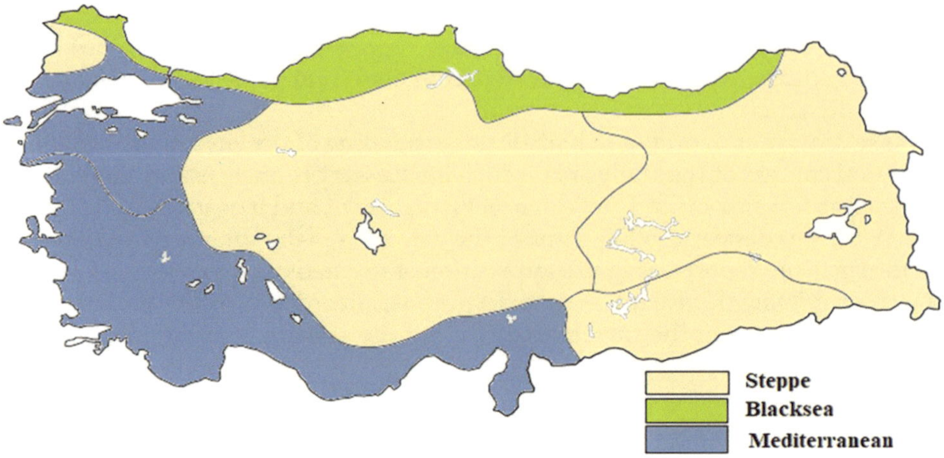

Figure 2.
Climate regions in Turkey [23].

by snow and frost. The region experiences significant daily and annual temperature fluctuations.

As shown in **Figure 3**, the average temperature in January, the coldest month, is −0.2°C, while in July, the hottest month, it rises to 23.1°C, with an annual average of 11.7°C. The number of rainy days ranges from 70 to 114, averaging 87 days per year. May, the wettest month, averages 10.4 rainy days.

Most precipitation occurs in winter and spring, with 34.87% of annual rainfall in spring, 31.62% in winter, 20.04% in autumn, and 13.47% in summer. The region mainly experiences convectional rainfall during the spring months. The annual average relative humidity is 63.7%, and the region enjoys an average of 7.04 hours of sunshine per day [25].

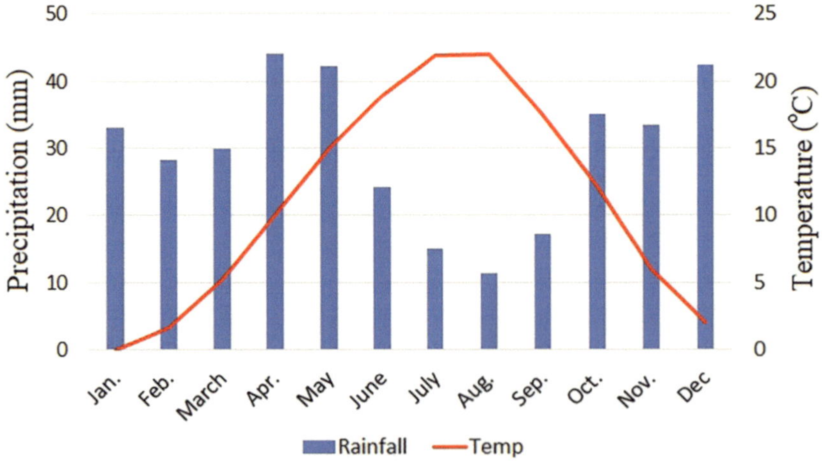

Figure 3.
Monthly temperature, precipitation in Central Anatolia [23].

2.3 Land use

The land use in the study area is presented in **Figure 4**, highlighting the distribution of different land categories such as agricultural areas, forests, pastures, and other uses.

In the study area, crops such as wheat, barley, oats, rye, chickpeas, safflower, lentils, and cumin are grown under dry conditions. Irrigated lands support the cultivation of additional crops like sugar beet, grain corn, silage corn, sunflower, potatoes, onions, dry beans, alfalfa, vetch, snack pumpkins, and snack sunflower [23].

Central Anatolia has a mix of coniferous and deciduous tree species. Coniferous trees include Taurus Cedar, Scots Pine, Black Pine, Red Pine, Cypress, and Arizona Cypress, while deciduous trees like *Robinia pseudoacacia*, *Ailanthus altissima*, Narrow-leaved Ash, Almond, and Oleaster are also found [26].

Pastures in Central Anatolia feature steppe vegetation, with annual plants dominating in low-altitude areas and perennial plants at higher altitudes. The region's pastures are largely in poor-to-moderate condition, with common plants including astragalus, Ivy-leaved speedwell, thyme, clover, sainfoin, and grasses like sheep's ruff and blue weed [27]. However, these plants are not abundant. Overgrazing, particularly in areas with intensive small livestock farming, has led to significant degradation of pasture quality. In some areas, plants like *Juncus effusus* and *Cynodon dactylon* dominate due to their resistance to grazing [28]. A study in Ankara province showed that only 60.55% of pasture area was covered by plants, with 39.45% as bare soil, indicating significant land degradation. The overall quality of pastures in Central Anatolia is rated as moderate to poor due to overgrazing, resulting in low-nutrient vegetation [29].

2.4 Soil

In the Central Anatolia region, 75.9% of the soils are characterized by clayey loam and clayey textures. The majority, 89.2%, of the soils are slightly alkaline, and nearly all (99.4%) are free from salinity. However, 85.5% of these soils exhibit low or very low organic matter content, and 56.1% are rich in calcium carbonate, indicating a high lime presence. Phosphorus availability is limited, with 75.4% of soils falling into medium, low, or very low categories. In contrast, 94.4% of soils contain high levels of available potassium. Similarly, 99.2% of the exchangeable calcium and 93.4% of the exchangeable magnesium levels are classified as sufficient, high, or very high. Micronutrient availability presents a challenge, as 44.8, 75.3, and 92.3% of soils show low or very low levels of available iron, zinc, and manganese, respectively. On the other hand, available copper is abundant, with 98.8% of soils classified as sufficient [30]. Organic carbon

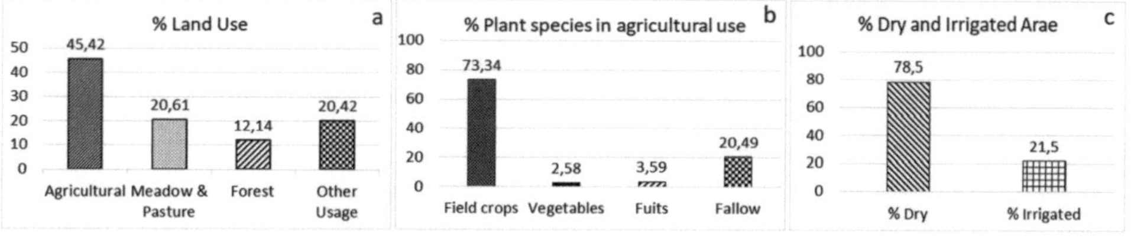

Figure 4.
Land use groups (a), plant species (b), and dry and irrigated area (c) in Central Anatolia [26].

is widely regarded as a vital indicator in the sustainable management of soils [31]. Regarding organic carbon levels, 18% of Turkish soils are classified as very low and 70% as low, and a total of 88% are considered problematic and insufficient [32].

2.5 Topography

Turkey's average elevation is 1141 meters, with a slope of 17%. In Central Anatolia, the average elevation is higher, at 1205 meters, with a slope of 9.6%. Areas below 500 meters make up 17.9% of Turkey's total area, while this proportion is only 0.1% in Central Anatolia. Elevations between 1000 and 1250 meters cover 38.3% of Central Anatolia, while areas above 1500 meters account for 17.5% of the region. About 6.2% of Turkey's land consists of areas with a slope of less than 1%, while 12.6% of the land has a slope of less than 2%. Regions with slopes greater than 20% make up 34% of the country. Meanwhile, areas with a slope of 2–5% account for 13.6%, and those with a slope between 5 and 20% cover about 40%. Central Anatolia is notable for having the largest proportion of land with a slope of less than 1%. The distribution of slopes in this region is as follows: 21.3% of the land has a slope of 2–5%, 22.8% has a slope of 5–10%, and 21% has a slope of 10–20% [33].

2.6 Tillage management

In Turkey, various soil processing tools are used for both dry and irrigated agriculture. The plow is the primary tool for initial soil processing in both systems. In dry farming, some fields are left fallow to conserve moisture due to insufficient rainfall. Primary soil processing is typically done in April or May using a moldboard or disc plow at a depth of 20–30 cm. Secondary tools such as cultivators or rotavators are used to level the soil, break up clods, and manage weeds, typically working at depths of 10–15 cm. A combination of cultivator and harrow may also be used for a third round to prepare the seedbed, with roller tools employed to smooth and press the soil before planting.

Tillage equipment such as tillers is crucial for preparing seedbeds, aerating the soil, managing weeds, and mixing soil. Around 50% of Turkey's soil has a heavy or medium-heavy texture [32], making it prone to compaction from heavy field traffic and monoculture practices. This compaction creates a "plow base" layer, restricting root development and reducing productivity. To address this, chisel plows and tools like rippers or subsoilers are recommended every 3–4 years to relieve compaction.

In irrigated farming, a wider variety of tools are used. For crops with late harvests, such as sugar beet and grain corn, a rotavator is often used after plowing for rapid seedbed preparation, particularly in dry conditions. Sometimes a second pass with the rotavator is necessary to break up large particles and ensure a smooth seedbed.

2.7 Irrigation management

Turkey has 24 million hectares of agricultural land, with 12 million hectares suitable for irrigation. Currently, 6.7 million hectares are irrigated, with 61.6% of the land using surface irrigation and 38.4% utilizing pressurized irrigation methods. In Central Anatolia, 22% of agricultural areas are irrigated.

DOI: http://dx.doi.org/10.5772/intechopen.1009836

However, several challenges affect irrigated agriculture in Turkey:

1. *Excessive and insufficient water use*: Farmers often misuse water, either overusing or underusing it, which reduces agricultural productivity and harms environmental resources. Excessive water use depletes water supplies, while insufficient use fails to meet crop needs, causing erosion, plant diseases, and environmental damage.

2. *Outdated and poorly maintained irrigation networks*: Aging irrigation systems lead to water losses and inefficient water use. These systems increase costs for farmers and contribute to environmental issues such as soil salinization.

3. *Water pollution*: Agricultural chemicals and untreated wastewater pollute irrigation water. Industrial discharges further contaminate water resources, harming soil and plants and leading to economic losses.

4. *Water transmission and distribution issues*: Inefficient water transmission systems cause significant water losses. Surface irrigation systems have an efficiency rate of 40–60%, while pressurized systems reach about 90% efficiency.

5. *Organizational and management challenges*: Poor coordination between farmers, irrigation unions, and government bodies leads to ineffective water management. Issues include inefficient irrigation scheduling, limited training, and lack of financial resources for system maintenance.

6. *Farmer education deficiencies*: Many farmers lack proper training, limiting their ability to adopt modern irrigation techniques, leading to productivity loss and environmental degradation.

7. *Monitoring and evaluation problems*: Ineffective monitoring of irrigation systems results in excessive water use, financial strain, and environmental degradation. Unregulated practices exacerbate water loss, erosion, and salinity, while uneven water distribution fuels social conflicts.

2.8 Soil erosion

Turkey's average elevation is 1132 meters, and 60% of the country has slopes greater than 12%. Slopes exceeding 20% cover 45% of the country, while 20% has slopes greater than 40%, and 6.5% exceeds 60%. Soil erosion is a significant issue, leading to an annual loss of 642 million tons of soil (8.24 tons per hectare) (**Figure 5**).

In Central Anatolia, the severity of erosion varies: 70.88% of the land experiences very light erosion, 13.87% light, 7.29% moderate, 4.53% severe, and 3.43% very severe erosion. Severe erosion affects 11.5% of areas with slopes above 12%, while only 1.3% of regions with slopes greater than 60% are impacted (**Figure 6**). Soil loss in Central Anatolia is influenced by several factors: topography (57.35%), vegetation (30.6%), precipitation (9.48%), and soil characteristics (2.57%). Of the total soil loss, 49.15% comes from agricultural lands, 46.07% from pasture areas, and 4.78% from forests. The average annual soil loss of 8.42 tons per hectare highlights the need for soil conservation measures. Soil loss is high in

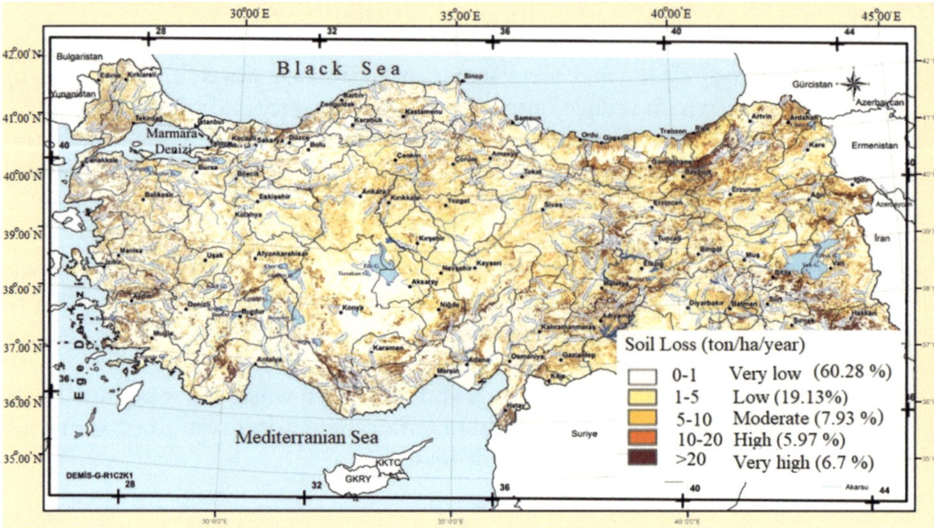

Figure 5.
Soil erosion map of Turkey [26].

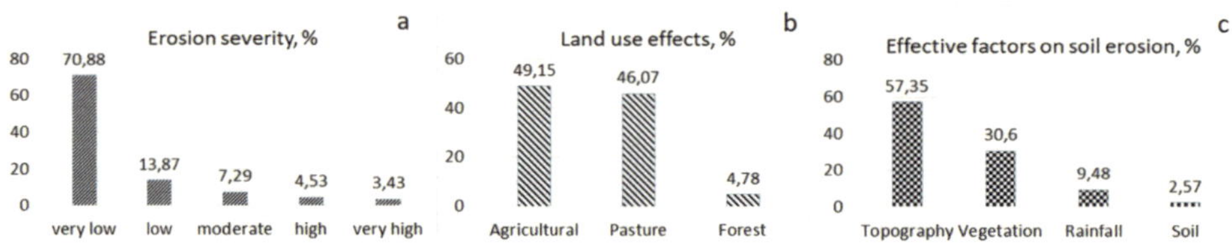

Figure 6.
Erosion severity (a), land use effects (b), effective factors on soil erosion (c) in Central Anatolia.

both agricultural and pasture lands. In pasture areas, the annual average soil loss is 18.36 tons per hectare, emphasizing the need for improved management strategies. Erosion severity in Central Anatolia is generally low, with agricultural lands contributing most to erosion, while forests contribute the least. The main factors driving erosion in the region are topography and vegetation [26].

3. Results

The Central Anatolia faces severe challenges including drought, water stress, and soil degradation. Insufficient rainfall and soils lacking organic matter and nutrients hinder plant growth. Around 75% of the soil is prone to compaction, affecting productivity. Key issues include:

1. *Tillage practices*: The overuse of plowing exacerbates moisture loss by disturbing the soil, especially in arid conditions.

2. *Monoculture farming*: Dominated by wheat and barley, monoculture farming depletes soil nutrients and increases pest vulnerability.

3. *Pasture quality*: Overgrazing has led to the degradation of pastures, with 30–35% being bare and over 90% classified as poor quality.

4. *Erosion and soil loss*: Agricultural lands and pastures are the main sources of erosion, worsened by poor vegetation cover and inadequate conservation practices.

5. *Water resources and irrigation*: Only 22% of agricultural land is irrigated, primarily using surface irrigation, leading to excessive water use and soil erosion.

6. *Groundwater depletion*: Unregulated groundwater pumping, especially in the Konya Plain, has caused sinkholes and reduced agricultural productivity.

7. *Water scarcity*: Water demand exceeds supply, affecting both agricultural and drinking water, and contributing to salinization.

8. *Climate change and forest health*: Decreased snowfall due to global warming threatens forests, requiring effective management.

9. *Land consolidation*: The lack of land consolidation results in fragmented agricultural practices, reducing productivity. Only half of Turkey's irrigable land is utilized effectively.

4. Sustainable soil and water management

Sustainable soil and water management aims to preserve soil health and fertility, prevent erosion, and ensure the efficient use of water. Conservation management includes various methods that improve soil structure, reduce water loss, and manage natural resources sustainably. To achieve these goals, the focus is on topics such as preventing soil erosion, preventing soil compaction, water management and irrigation techniques, improving soil structure, mulching-green manuring, supporting biodiversity, preventing soil compaction and transport, and soil monitoring and management systems.

Sustainable land and water management plays an active role in improving environmental quality by fulfilling numerous functions in achieving targeted objectives: (1) reducing surface runoff, protecting soil, and preventing erosion, (2) improving land use, (3) adopting best management practices, (4) improving water quality, and (5) preserving biodiversity. To achieve these goals, it employs land use regulation, physical soil conservation measures (terracing and windbreaks), biological soil conservation measures (cover crops, crop rotation, green manuring, conservation tillage, contour farming, strip cropping, mulching, and tillage management), and chemical soil conservation methods. These methods are applied integrally in sustainable pasture and water management.

4.1 Physical soil conservation methods

Mechanical and engineering methods are used to prevent erosion [34] and soil degradation by slowing water flow [35], increasing infiltration [36], and reducing runoff [37] and these methods improve soil stability and fertility [38]. In Central

Anatolia, agricultural lands with slopes of 3 to 4.5% are suitable for these erosion control techniques, boosting productivity.

4.1.1 Terracing

Terraces are step-like structures built on slopes to reduce soil erosion [39] and water runoff while improving water infiltration [40]. Their dimensions depend on factors like slope, soil, rainfall, and vegetation. Drainage systems prevent water accumulation [41], while vegetation stabilizes terraces, secures soil, and helps slow water absorption, reducing erosion risks (**Figure** 7).

Figure 7.
Terracing works on sloping areas [26].

4.1.2 Windbreaks

Trees and shrubs play a vital role in reducing wind erosion and preserving soil moisture [42]. Windbreaks, made up of rows of trees or shrubs, protect soil (**Figure 8**), control erosion [43], retain moisture, and create microclimates [44]. They also prevent plant damage [45], support biodiversity [46], and slow wind speed [42]. During critical growth stages, they help retain moisture and prevent snow cover loss [47]. To maximize benefits, windbreaks should be strategically planted and maintained, using species adapted to local climate and soil conditions.

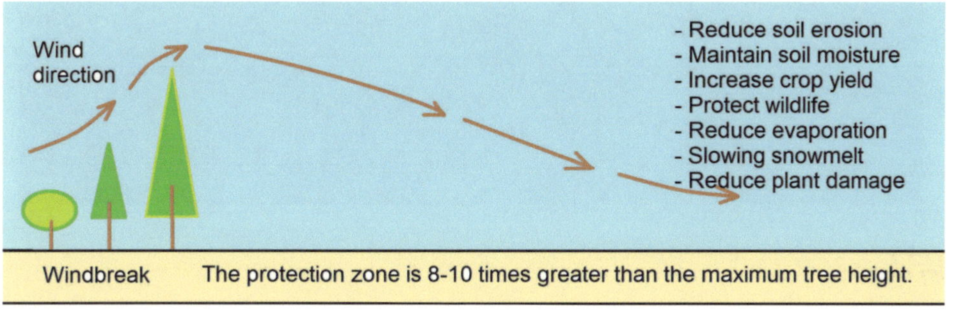

Figure 8.
Windbreak and its effects.

4.2 Biological soil conservation methods

These methods use natural processes and plants to protect soil health, relying on living organisms instead of chemical inputs. These eco-friendly methods prevent erosion, improve water infiltration, and stabilize soil through vegetation. Compared to bare soil, vegetated areas face less erosion. Biological practices, often combined with other techniques, enhance soil fertility, moisture retention, and biodiversity. They are cost-effective and support sustainable farming by preserving soil health and sequestering carbon.

4.2.1 Cover cropping

Cover crops have several functions such as reducing evaporation [48], maintaining soil moisture [49, 50], conserving nutrient and water, improving soil quality, reducing the risk of soil erosion and weed control [51], managing soil pests depending on environmental factors [52], increasing the effectiveness of N fertilizer [53], reducing nitrogen loss by reducing soil erosion, and, in certain species, reducing the C:N ratio of crop residues [54], providing an important source of organic matter to the soil [55], nutrient cycling, aggregate stability, etc. It has roles such as helping to form soil [56], improving soil structure [57], and reducing nitrate leaching in soil (**Figure 9**) [58].

Figure 9.
Cover cropping.

4.2.2 Crop rotation

It is an important practice that ensures the sustainability of agricultural production by improving the physical, chemical, and biological properties of the soil. This method increases productivity while maintaining soil health and minimizing environmental impacts. Crop rotation is the practice of growing different plants in a specific sequence on the same field. This method is used to maintain soil fertility [59], reduce diseases and pests [60], improve soil structure [61], and ensure the sustainability of agricultural production. Growing plants with varying root depths through crop rotation positively impacts soil physical structure, increases the soil's resistance to erosion [62], and enhances its water retention capacity. Different plants take up and release different nutrients from the soil. For example, legumes fix nitrogen, enriching the soil with nitrogen, while cereals consume nitrogen. This cycle helps maintain the nutrient balance in the soil. The incorporation of plant residues into the soil increases the organic matter content, which improves the chemical structure of the soil and accelerates the nutrient cycle [63]. Different plants support different populations of microorganisms, enhancing the biological diversity and microbial activity in the soil [64], which in turn improves soil health [65]. Additionally, the differences between plants help break the developmental cycles of specific diseases and pests, keeping their populations under control [66, 67].

In arid regions, the focus is on maximizing soil moisture retention and selecting crops that utilize water stored during rainy seasons, like wheat, barley, and legumes. Crops with high water demands, such as sunflowers and sugar beets, should be avoided. This practice enhances soil health and stabilizes yields, ultimately reducing costs.

Cereals and legumes: A typical rotation alternates cereals (e.g., wheat, barley, oats) with legumes (e.g., lentils, chickpeas, peas). Legumes fix nitrogen, boosting soil fertility for the next cereal crop.

Cereals and fallow: In low-rainfall areas, cereal crops are often followed by a fallow period to replenish soil moisture and reduce degradation risks.

Diverse crop mixes: Including crops like canola or forage plants can disrupt pest cycles, enhance soil structure, and improve overall soil health.

4.2.3 Green manuring

This is used to improve soil health [68], accumulate nitrogen [69], ensure environmental sustainability [70], improve water holding capacity [71], reduce the risk of crop diseases [72], mitigate the adverse effects of climate change [73], contribute to a healthier and more resilient ecosystem by increasing biodiversity and supporting the activity of beneficial microorganisms and macroorganisms in the soil [74], reduce the impact of soil erosion [75], reduce the residual effect of chemical fertilizers and pesticides [76], which is an agricultural practice that involves the cultivation of certain plants such as vetch, clover, and mustard to reduce soil compaction, make the soil more arable, aerate the soil [77], reduce the percentage of immobilization of various nutrients in the soil [78], increase the supply and uptake of plant nutrients [79], and improve the physico-chemical properties of the soil (**Figure 10**) [80].

These plants are grown or incorporated into the soil or left on the surface. As they decompose, they add organic matter [81], improve soil structure [82], fertility [83], and moisture retention [80], and fix nutrients such as nitrogen [77]. Green manuring also suppresses weeds [80], prevents erosion, and can disrupt pest cycles, reducing the use of chemical fertilizers [79]. This sustainable practice supports higher yields without synthetic fertilizers and improves the organic content of the soil, especially in the irrigated areas, increasing productivity for subsequent crops.

Figure 10.
Green manuring provides organic matter accumulation to the soil.

Figure 11.
Conservation tillage.

4.2.4 Conservation tillage (CT)

This practice reduces soil disturbance by minimizing the frequency and intensity of tillage. The core principle of CT is to minimize soil disturbance, which improves water storage and moisture retention, while also reducing soil erosion and the reliance on nitrogen fertilizers by increasing surface cover [84]. It helps maintain soil structure [85], retain moisture [86], control erosion, and increase soil porosity [87]. This method improves soil health, reduces labor and energy costs, and contributes to carbon storage [88], and increases soil fertility (**Figure 11**) [89].

Conservation tillage practices result in a decrease in bulk density, an increase in water holding capacity, and improved biological and physical soil properties through microbial activity [90]. Minimizing mechanical tillage practices helps decrease energy consumption and carbon emissions from fossil fuel use [91], along with reducing labor demands [92]. It boosts soil biodiversity and ecosystem functions, which can enhance agricultural productivity by promoting the accumulation of soil organic carbon, nutrient cycling, soil structure development, and crop resilience to pests and diseases [93]. Additionally, it acts as the primary exchange and storage reservoir for soil nitrogen [94]. Conservation tillage minimizes soil disturbance, increases soil organic matter over time, improves soil infiltration, and supports the formation of soil microorganisms. Conservation tillage, which includes reduced and zero tillage methods, helps decrease erosion and conserve soil moisture by keeping at least 30% of the soil surface covered. It is important for CT to remain a tillage practice that ensures a minimum of 30% of the soil's surface is covered with crop residues after planting the subsequent crop [95, 96]. Crop residues play an important role as a valuable and renewable resource [97]. CT affects the physical properties of the soil in long-term application [97], improves the subsoil environment [98], stabilizes soil temperature [99], increases soil porosity and water infiltration rate during heavy rains [100], improves aggregate stability limiting soil erosion [101], and reduces runoff and soil erosion [102]. However, these benefits may vary according to cropping system, climate, and soil type [103]. The addition of organic matter parallel to the decrease in the bulk density of the soil increases the porosity and makes it more resistant to degradation [104]. The classical practice of minimum tillage for a long period increases the soil organic matter content [105].

4.2.4.1 Reduced tillage

Reducing tillage provides several benefits, including less soil erosion [106], improved water retention [107], increased organic matter [108], and reduced fuel consumption and carbon emissions [109]. Practices like leaving crop residues on the soil surface and direct seeding without tilling preserve soil structure and organic matter, enhance water infiltration [110], and reduce evaporation [111]. These approaches also lower labor and fuel costs, promote carbon storage in soil [112], and support climate change mitigation [113], making reduced tillage a sustainable agricultural practice that enhances productivity [114] and minimizes environmental impacts (**Figure 12**).

Figure 12.
Reduced tillage.

4.2.4.2 No-till farming

No-till farming enhances soil health [115] and biodiversity by minimizing soil disturbance. Seeds are directly planted without plowing, and crop residues are left on the soil to protect and enrich it with organic matter [116]. This practice supports environmental sustainability [117] and offers economic benefits by preserving soil structure [118] and biodiversity (**Figure 13**) [119].

Figure 13.
No tillage farming in soybean.

4.2.5 Contour farming

Contour farming involves planting crops parallel to land contours to reduce surface runoff and soil erosion [120, 121], retaining rainfall, and maintaining fertility [122]. This method can decrease soil erosion by up to 50% and is effective on slopes of 2–10%. It slows water flow [123], increases infiltration [124], and is even more effective when combined with other conservation practices like zero-till or minimum-till farming (**Figure 14**).

Figure 14.
Contour farming.

4.2.6 Strip cropping

Strip cropping involves alternating strips of crops with different growth and harvest times along land contours to reduce soil erosion [37] and improve water infiltration [125]. It slows surface runoff [126], preserves soil fertility, and disrupts pest cycles, reducing the need for pesticides [127]. Studies show strip cropping can increase yields by up to 59% and reduce soil loss by 18%. It also conserves water and boosts biomass, enhancing both productivity and sustainability (**Figure 15**) [128].

Figure 15.
Strip cropping.

4.2.7 Mulching

Mulching is a water-saving technique used in arid regions to conserve soil moisture and reduce evaporation [129]. It involves covering the soil with materials like plastic or wheat straw to minimize moisture loss [130] and control weeds [131]. This method is particularly effective in vegetable and fruit farming, reducing the need for chemical herbicides [132] while retaining soil moisture for longer periods (**Figure 16**).

Figure 16.
Mulching.

4.2.8 Tillage management

Effective tillage management in dryland areas focuses on conserving soil moisture [133], reducing erosion [134], and improving soil health [135]. In arid climates, selecting the right tillage tools is essential to minimize soil disturbance [136], preserve moisture, manage crop residues [137], and optimize fuel and labor usage for cost-effectiveness [138]. The appropriate tools can significantly support these objectives, ensuring better crop production [67] and soil preservation [139]. Some recommended tillage tools are knife rollers, chisel plows, disk harrows, subsoilers, field cultivators.

4.3 Chemical methods in soil conservation

In climate regions dominated by drought, the amount of plant nutrients and organic matter in the soil is insufficient. Sustainability of agricultural production requires the use of sufficient amounts of chemical fertilizers [140], at appropriate times and the use of organic matter [141]. It also involves improving soil properties with soil conditioners [142], reducing plant nutrient uptake due to high lime and pH levels [143], salt management [144], erosion control [145], pesticide and herbicide use [146], improving soil pollution caused by chemicals [147], conserving underground and surface water resources [148], integrated nutrient management [149], promoting organic farming [150], farmer education and awareness [151], use of soil enhancers [152], regulating soil pH [153], effective use of nutrients [154], and fertilization management [155] can be focused on to achieve these goals.

5. Rangeland and pasture management in arid climates

Rangelands are important and complementary components of sustainable natural resource management. In arid climate regions, the management of rangelands is related to the continuity of natural resources within the ecological balance. Solving existing problems requires the presence of multiple perspectives. Under these conditions, key issues in rangeland management include water management [156], vegetation management [157], grazing management [158], soil health [159], climate change adaptation [160], biodiversity enhancement [161], erosion control [162], and farmer education [163].

6. Water use and management in arid climates

Water use and management are essential components for sustainably utilizing land and water resources, along with ensuring food security for the growing global population. In irrigated lands, limited irrigation water applications [164], development of effective water management strategies [165], monitoring soil moisture in the root zone during irrigation and applying techniques based on this [166], improving irrigation during critical plant growth periods [167], the management of underground and surface water resources [168], water management techniques under drought conditions [169], the use of wastewater in irrigation [170], organization and management of water transmission and distribution systems [171], monitoring and evaluation systems in irrigation [172], improvement of farmers' irrigation practices [173], excessive water use [174], adaptation of modern irrigation systems [175], use of saline and low-quality waters [176], treatment, and use of domestic and industrial wastewater [177] are just some of the many diverse topics that need to be evaluated in a sustainable and integrated manner.

In Central Anatolia, challenges like excessive water use, poor water quality, and inefficient irrigation systems hinder sustainable water management. Proposed solutions include integrating field development with irrigation systems, adopting pressurized conveyance systems, and transitioning to more efficient irrigation methods. Key strategies include educating farmers on water management, implementing soil moisture monitoring technologies, and creating crop rotation programs. Deficit irrigation tailored to plant growth stages, fertigation practices, and legal measures to protect water resources will further optimize water use. These actions aim to improve efficiency, reduce waste, and ensure sustainable water resources.

7. Conclusion

This chapter provides a summary of soil and water conservation methods applicable to Central Anatolia and similar arid and semi-arid climate regions. Issues encountered in such climates due to global warming often share commonalities. While regional challenges in drought-affected areas may vary, combining physical, chemical, and biological approaches, as outlined here, can lead to more efficient and effective solutions. Therefore, there is no single remedy for the environmental problems caused by drought. A sustainable and healthy environment is only achievable through the preservation of its constituent elements.

Author details

Ertuğrul Karaş
Faculty of Agriculture, Department of Soil Science and Plant Nutrient,
Eskişehir Osmangazi University, Turkey

*Address all correspondence to: ertugrulkaras@gmail.com

IntechOpen

© 2025 The Author(s). Licensee IntechOpen. This chapter is distributed under the terms of the Creative Commons Attribution License (http://creativecommons.org/licenses/by/4.0), which permits unrestricted use, distribution, and reproduction in any medium, provided the original work is properly cited.

References

[1] Seppelt R, Klotz S, Peiter E, Volk M. Agriculture and food security under a changing climate: An underestimated challenge. iScience. 2022;**25**:105551. DOI: 10.1016/j.isci.2022.105551

[2] UNICEF. Water Scarcity. Available from: https://www.unicef.org/wash/water-scarcity [Accessed: December 15, 2024]

[3] Meier J, Zabel F, Mauser W. A global approach to estimate irrigated areas – A comparison between different data and statistics. Hydrology and Earth System Sciences. 2018;**22**:1119-1133. DOI: 10.5194/hess-22-1119-2018

[4] Mallareddy M, Thirumalaikumar R, Balasubramanian P, Naseeruddin R, Nithya N, Mariadoss A, et al. Maximizing water use efficiency in Rice farming: A comprehensive review of innovative irrigation management technologies. Water. 2023;**15**(10):1802. DOI: 10.3390/w15101802

[5] Hui Q, Tianjun Z, Liwei Z, Jie J, Xiaolong C, Shuai H. Future changes in precipitation and water availability over the Tibetan plateau projected by CMIP6 models constrained by climate sensitivity. Atmospheric and Oceanic Science Letters. 2024;**17**(5):100537. DOI: 10.1016/j.aosl.2024.100537. ISSN 1674-2834

[6] UN. World Population. Available from: https://www.un.org/en/global-İssues/population#:~:text=The%20world%20population%20is%20projected,and%2010.4%20billion%20by%202100 [Accessed: December 15, 2024]

[7] Pedro MM, Soares D, Lima CA. Water scarcity down to earth surface in a Mediterranean climate: The extreme future of soil moisture in Portugal. Journal of Hydrology. 2022;**615**(Part B):128731. DOI: 10.1016/j.jhydrol.2022.128731. ISSN 0022-1694

[8] Xoplaki E, González-Rouco F, Luterbacher J, Wanner H. Wet season Mediterranean precipitation variability: Influence of large-scale dynamics and trends. Climate Dynamics. 2004;**23**:63-78

[9] Solomon SD, Qin M, Manning Z, Chen M, Marquis KB, Averyt M, et al., editors. IPCC Climate Change 2007: The Physical Science Basis. Contribution of Working Group I to the Fourth Assessment Report of the Intergovernmental Panel on Climate Change. Cambridge, United Kingdom: Cambridge University Press; 2007

[10] Milano M, Ruelland D, Fernandez S. Facing climatic and anthropogenic changes in the Mediterranean basin: What will be the medium-term impact on water stress? Comptes Rendus Geoscience. 2012;**344**:432-440

[11] Milano M, Ruelland D, Fernandez S, Dezetter A, Fabre J, Servat E, et al. Current state of Mediterranean water resources and future trends under climatic and anthropogenic changes. Hydrological Sciences Journal. 2013;**58**(3):498-518. DOI: 10.1080/02626667.2013.774458

[12] Mediterranean Water Scarcity & Drought Working Group (MED WS&D WG) (2007-04-00) European Commission-Environment. Available from: http://www.emwis.net/topics/WaterScarcity [Accessed: December 15, 2024]

[13] UNDP. Climate Change & Turkey Impacts, Sectoral Analyses,

Socio-Economic Dimensions. United Nations Development Programme Turkey Office. Available from: https://www.undp.org/sites/g/files/zskgke326/files/migration/tr/Report.pdf [Accessed: December 15, 2024]

[14] TUIK. Türkiye Population Statistics Bulletin (in Turkish). Available from: https://data.tuik.gov.tr/ [Accessed: December 15, 2024]

[15] Anonymous. General agricultural sector report (in Turkish). Available from: https://ticaret.gov.tr/data/ [Accessed: December 15, 2024]

[16] Polat K. Wheat (in Turkish), TEPGE Agricultural Economics and Policy Development Institute/AEPDI, Available from: https://arastirma.tarimorman.gov.tr/tepge/Belgeler/ [Accessed in February 12, 22024]

[17] Asik G, Karakoç U, Pamuk S. Regional inequalities and the west–east divide in Turkey since 1913. The Economic History Review. 2023;**76**:1305-1332. DOI: 10.1111/ehr.13245

[18] UNDP Climate Change Adaptation Report. 2022. Climate Adaptation Action is Enhancing in Turkey. 30 May 2022. Available from: https://www.undp.org/turkiye/news/climate-adaptation-action-enhancing-turkey [Accessed: December 15, 2024]

[19] Karas E. The Effects of Climate Change on Stream Flow. Rural Research Institute of Eskisehir; 2006. p. 11-12. (in Turkish, Unpublished Report)

[20] Kesici E. Lakes in Turkey are Drying Up. 2024. Available from: https://www.tridge.com/news/lakes-in-turkey-are-drying-up-kwbqla [Accessed: December 15, 2024]

[21] Gür H. Global climate change and biodiversity (in Turkish). 2016. https://www.researchgate.net/publication/308917834_Kuresel_iklim_degisikligi_ve_biyolojik_cesitlilik [Accessed: December 15, 2024]

[22] Vikipedia. Central Anatolia. https://en.wikipedia.org/wiki/Central_Anatolia_region [Accessed: December 15, 2024]

[23] Vikipedia. Geographic Regions of Turkey (in Turkish). https://tr.wikipedia.org/wiki/T%C3%BCrkiye%27nin_co%C4%9Frafi_b%C3%B6lgeleri [Accessed: December 15, 2024]

[24] Sensoy S, Demircan M, Ulupınar Y, Balta İ. Climate of Turkey, General Directorate of State Meteorological Affairs. Available from: https://mgm.gov.tr/FILES/iklim/yayinlar/turkiye_iklimi.pdf [Accessed: December 15, 2024]

[25] Climate Data in Turkey (in Turkish) General Directorate of State Meteorological Affairs 2024. https://mgm.gov.tr [Accessed: December 15, 2024]

[26] Kokaş Y, Çetin B, Sargıncı M, Ayyüce F. Afforestation works in Central Anatolia (in Turkish). Forestry Journal. 2018;**14**(1):1-20

[27] Avağ A, Koç A, Kendir H. National Rangeland use and Management Project (in Turkish). Ankara: Tübitak Kamag, Project Report no 106G017, 1226 s; 2012

[28] Karahan AD, Başbağ M. Determination of botanical composition, pasture quality degree and pasture conditions of different aspects of Savucak Mountain pastures (Diyarbakır-Çüngüş) (in Turkish). ISPEC Journal of Agricultural Sciences. 2023;7(2):375-386. DOI: 10.5281/zenodo.80 50963

[29] Unal S, Mutlu Z, Mermer A, Urla Ö, Ünal E, Aydoğdu M, et al. A study on the evaluation of pastures in Ankara

province (in Turkish). Tarla Bitkileri Merkez Araştırma Enstitüsü Dergisi. 2012;**21**(2):41-49

[30] Akın A, Taşova H. Determination and mapping of some productivity parameters of agricultural soils in the Central Anatolia region (in Turkish). Mediterranean Agricultural Sciences. 2019;**32**:1-6

[31] Koorneef GJ, Pulleman MM, Comans RNJ, van Rijssel SQ, Barré P, Baudin F, et al. Assessing soil functioning: What is the added value of soil organic carbon quality measurements alongside total organic carbon content? Soil Biology and Biochemistry. 2024;**196**:109507. DOI: 10.1016/j.soilbio.2024.109507, ISSN 0038-0717

[32] Anonymous. Creation of a Geographical Database of the Productivity and Organic Carbon Content of some of the Soils in Türkiye (in Turkish). Ankara, Turkey: Ministry Of Agrıculture And Forestry, General Directorate of Agricultural Research and Policies, Result report, TAGEM/TSKAD/11/A13/P03; 2018, 86 pages

[33] Elibüyük M, Yılmaz E. Elevation steps and slope groups according to geographic regions and sections of Turkey (in Turkish). Coğrafi Bilimler Dergisi. 2010;**8**(1):27-56. DOI: 10.1501/Cogbil_0000000104

[34] Zhang QQ, Chen ZQ, Chen ZB, Ma QY. Characteristics of slope runoff and sediment yield and Rees migration under different soil and water conservation measures in a typical red soil region of southern China. Journal of Soil and Water Conservation. 2018;**32**:46-53

[35] Chen X, Liang Z, Zhang Z, Zhang L. Effects of soil and water conservation measures on runoff and sediment yield in red soil slope farmland under natural rainfall. Sustainability. 2019;**12**(8):3417. DOI: 10.3390/su12083417

[36] Thierfelder C, Wall PC. Effects of conservation agriculture techniques on infiltration and soil water content in Zambia and Zimbabwe. Soil and Tillage Research. 2009;**105**(2):217-227. DOI: 10.1016/j.still.2009.07.007

[37] Rajbanshi J, Das S, Paul R. Quantification of the effects of conservation practices on surface runoff and soil erosion in croplands and their trade-off: A meta-analysis. Science of the Total Environment. 2023;**864**:161015. DOI: 10.1016/j.scitotenv.2022.161015

[38] Sinore T, Kissi E, Aticho A. The effects of biological soil conservation practices and community perception toward these practices in the Lemo District of southern Ethiopia. International Soil and Water Conservation Research. 2018;**6**(2):123-130. DOI: 10.1016/j.iswcr.2018.01.004

[39] Vanmaercke M, Poesen J, Maetens W, de Vente J, Verstraeten G. Sediment yield as a desertification risk indicator. Science of the Total Environment. 2011;**409**(9):1715-1725

[40] Chen D, Wei W, Chen L. Effects of terracing practices on water erosion control in China: A meta-analysis. Earth-Science Reviews. 2017;**173**:109-121. DOI: 10.1016/j.earscirev.2017.08.007

[41] Deng C, Zhang G, Liu Y, Nie X, Li Z, Liu J, et al. Advantages and disadvantages of terracing: A comprehensive review. International Soil and Water Conservation Research. 2021;**9**(3):344-359. DOI: 10.1016/j.iswcr.2021.03.002

[42] Řeháček D, Khel T, Kučera J, Vopravıl J, Petera M. Effect of windbreaks on wind speed reduction and soil protection against wind erosion. Soil and Water Research. 2017;**12**(2):128-135. DOI: 10.17221/45/2016-SWR

[43] Chang X, Sun L, Yu X, Liu Z, Jia G, Wang Y, et al. Windbreak efficiency in controlling wind erosion and particulate matter concentrations from farmlands. Agriculture, Ecosystems & Environment. 2021;**308**:107269. DOI: 10.1016/j.agee.2020.107269

[44] Baker T, Moroni M, Hunt M, Worledge D, Mendham D. Temporal, environmental and spatial changes in the effect of windbreaks on pasture microclimate. Agricultural and Forest Meteorology. 2021;**297**. DOI: 10.1016/j.agrformet.2020.108265

[45] Cleugh H. Effects of windbreaks on airflow, microclimates and crop yields. Agroforestry Systems. 1998;**41**:55-84. DOI: 10.1023/A:1006019805109

[46] Podhrázská J, Kučera J, Doubrava D, Doležal P. Functions of windbreaks in the landscape ecological network and methods of their evaluation. Forests. 2021;**12**(1):67. DOI: 10.3390/f12010067

[47] Brandle J, Nickerson H. EC96-1770 Windbreaks for Snow Management. Historical Materials from University of Nebraska-Lincoln Extension. 1996:844. Available from: https://digitalcommons.unl.edu/extensionhist/844

[48] Sharma P, Singh A, Kahlon CS, Brar AS, Grover KK, Dia M, et al. The role of cover crops towards sustainable soil health and agriculture: A review paper. American Journal of Plant Sciences. 2018;**9**:1935

[49] Nielsen DC, Lyon DJ, Hergert GW, Higgins RK, Calderon FJ, Vigil MF. Cover crop mixtures do not use water differently than single-species plantings. Agronomy Journal. 2020;**107**:1025-1038

[50] Blanco-Canqui H, Ruis SJ. Cover crop impacts on soil physical properties: A review. Soil Science Society of America Journal. 2020, 2020;**84**:1527-1576. DOI: 10.1002/saj2.20129

[51] Teasdale JR, Daughtry CST. Weed suppression by live and desiccated hairy vetch (Vicia villosa). Weed Science. 1993;**41**:207-212

[52] White CA, Holmes HF, Morris NL, Stobart RM. Cover crops: A review of the benefits, management practices and knowledge gaps. Research Review No. 90 A review of the benefits, optimal crop management practices and knowledge gaps associated with different cover crop species, In 14th ESA Congress 5e9th September 2016. Edinburgh: Scotland 2016

[53] Gabriel JL, Alonso-Ayuso M, García-Gonzalez L, Hontoria C, Quemada M. Nitrogen use efficiency and fertilizer fate in a long-term experiment with winter cover crops. European Journal of Agronomy. 2016;**79**:14-e22

[54] Dabney SM, Delgado JA, Meisinger JJ, Schomberg HH, Liebig MA, Kaspar T. Using cover crops and cropping systems for nitrogen management. In: Delgado JA, Follett RF, editors. Advances in Nitrogen Management for Water Quality. Ankeny, IA: Soil Water Conservation Society; 2010. pp. 231-e282

[55] Sainju UM, Whitehead WF, Singh BP. Biculture legume cereal cover crops for enhanced biomass yield and carbon and nitrogen. Agronomy Journal. 2005;**97**:1403-e1412

[56] Norris CE, Congreves KA. Alternative management practices improve soil health indices in intensive

vegetable cropping systems: A review. Frontier Environment Science. 2018;**6**:50. DOI: 10.3389/fenvs.2018.00050

[57] Mcgourty G, Reganold JP. Managing vineyard soil organic matter with cover crops. In: Proceedings of the Soil Environment and Vine Mineral Nutrition Nutrition Symposium. Christensen P, Smart DR, editors. Davis, CA: American Society for Enology and Viticulture; 2005. pp. 145-151

[58] Ruffo ML, Bullock DG, Bollero GA. Soybean yield as affected by biomass and nitrogen uptake of cereal rye in winter cover crop rotations. Agronomy Journal. 2004;**96**:800-805

[59] Yang X, Xiong J, Du T, Ju X, Gan Y, Li S, et al. Diversifying crop rotation increases food production, reduces net greenhouse gas emissions and improves soil health. Nature Communications. 2024;**15**(1):1-14. DOI: 10.1038/s41467-023-44464-9

[60] Pandey S. Crop rotation and intercropping techniques. In: Modern Concept in Agronomy. ND Global Publication House; 2024. DOI: 10.5281/zenodo.11051201. ISBN: 978-81-972002-0-5

[61] Liu X, Wu X, Liang G, Zheng F, Zhang M, Li S. A global meta-analysis of the impacts of no-tillage on soil aggregation and aggregate-associated organic carbon. Land Degradation & Development. 2021;**32**:5292-5305. DOI: 10.1002/ldr.4109

[62] Li C, Shi W, Huang M. Effects of crop rotation and topography on soil erosion and nutrient loss under natural rainfall conditions on the Chinese loess plateau. Land. 2023;**12**(2):265. DOI: 10.3390/land12020265

[63] German RN, Thompson CE, Benton TG. Relationships among multiple aspects of agriculture's environmental impact and productivity: A meta-analysis to guide sustainable agriculture. Biological Reviews. 2017;**92**:716-738

[64] Liang C, Schimel JP, Jastrow JD. The importance of anabolism in microbial control over soil carbon storage. Nature Microbiology. 2017;**2**:17105. DOI: 10.1038/nmicrobiol.2017.105

[65] Zang H, Blagodatskaya E, Wen Y, Xu X, Dyckmans J, Kuzyakov Y. Carbon sequestration and turnover in soil under the energy crop Miscanthus: Repeated 13C natural abundance approach and literature synthesis. GCB Bioenergy. 2018;**10**:262-271. DOI: 10.1111/gcbb.12485

[66] Schmer MR, Jin VL, Wienhold BJ, Becker SM, Varvel GE. Long-term rotation diversity and nitrogen effects on soil organic carbon and nitrogen stocks. Agrosystems, Geosciences & Environment. 2020;**3**:1-9. DOI: 10.1002/agg2.20055

[67] Zohry A, Ouda S. Crop Rotation Defeats Pests and Weeds: An Approach to Secure Future Food. Ouda S, Zohry AH, Noreldin T, editors. Giza: Springer; 2018. pp. 77-88. DOI: 10.1007/978-3-030-05351-2_5

[68] Pandey V, Srivastava A. Effect of integrated use of organic manures and chemical fertilizers under soil test crop response approach on soil properties and yield of maize (*Zea mays* L.). International Journal of Plant & Soil Science. 2021;**33**(10):40-47

[69] Irin IJ, Biswas P. Performance of different green manuring crops in Bangladesh. Research in Agriculture Livestock and Fisheries. 2021;**8**(1):25-31

[70] Javanmard A, Machiani MA, Haghaninia M, Pistelli L, Najar B.

Effects of green manures (in the form of monoculture and intercropping), biofertilizer and organic manure on the productivity and phytochemical properties of peppermint (*Menthe piperita* L.). Plants. 2022;**11**(21):2941

[71] Isah AS, Amans EB, Odion EC, Yusuf AA. Growth rate and yield of two tomato varieties (*Lycopersicon esculentum* mill) under green manure and nap fertilizer rate samaru northern Guinea savanna. International Journal of Agronomy. 2014:1-8. DOI: 10.1155/2014/932759

[72] Woźniak M, Gałązka A, Grządziel J, Frąc M. Microbial diversity of paulownia spp. leaves – A new source of green manure. Bio Resources. 2018;**13**(3):4807-4819

[73] Fabunmi TO, Adigbo SO, Odedina JN, Olasunkanmi TO. Effects of planting dates on green manure of cowpea (*Vigna unguiculata* L.), response of succeeding maize in a derived savanna ecological zone of Nigeria. Journal of Agricultural Science. 2012;**4**(7):1-57. DOI: 10.5539/jas.v4n7p57

[74] Auld DL, Bettis BL, Dial MJ, Murray GA. Austrian winter and spring peas as green manure crops in Northern Idaho. Agronomy Journal. 1982;**74**(6):1047-1050

[75] Bhayal D, Khaddar VK, Bhayal L, Yadav T, Bangar KS, Singh B. Effect of Sunhemp green manuring and intercropping on soil properties. International Journal of Current Microbiology and Applied Sciences. 2018;7:371-384. DOI: 10.20546/ijcmas.2018.712.046

[76] Billah S, Joarder J, Papy F. Amelioration of the Chemical Properties of Soil through Application of Green Manure Crops in Bajoa Soil Series. Khulna University Studies; 2022. pp. 211-216

[77] Moraes EC, Lima JS, Neto FB, Linhares PCA, Costa AP, Crispim JF, et al. Effects of different rooster tree (*Calotropisprocera*) amounts and spatial arrangements on the performance of the beet-cowpea intercropping system. Australian Journal of Crop Science. 2019;**13**(04):486-493

[78] Dinesh R, Dubey R. Nitrogen mineralization rates and kinetics in soils freshly amended with green manures. Journal of Agronomy and Crop Science. 1998;**181**(1):49-53

[79] Asghar W, Kataoka R. Green manure incorporation accelerates enzyme activity, plant growth, and changes in the fungal community of soil. Archives of Microbiology. 2021;**204**(1)

[80] Verma NS, Yadav D, Chouhan M, Bhagat C, Kochale P. Understanding potential ımpact of green manuring on crop and soil: A comprehensive review. Biological Forum – An International Journal. 2023;**15**(10):832-839

[81] Irin IJ, Biswas PK, Khan MA. Efficacy of different green manuring crops to soil fertility, yield and seed quality of T. Aman rice. Asian Journal of Medical and Biological Research. 2021;**7**(4):298-311

[82] Chandrasoorian S, Venkataraman NS, Babu R, Amutha R. Influence of organic and inorganic nutrients on yield, physicochemical properties and cooking quality of seeragasamba rice. International Journal of Chemical Studies. 2020;**8**(1):2780-2784

[83] Brito LF, Galvão JCC, Giehl J, Coellho SP, Campos SA, Barrella TP, et al. Decomposition of cover crop mulch and weed control under a no-till system

for organic maize. Bioscience Journal. 2019;**35**(5):1339-1348

[84] Gao WS. Development trends and basic principles of conservation tillage. Scientia Agricultura Sinica. 2007;**40**:2702-2708

[85] Busari MA, Kuka SS, Kaur A, Bhatt R, Dulazi AM. Conservation tillage impacts on soil, crop and the environment. International Soil and Water Conservation Research. 2015;**3**(2):119-129. DOI: 10.1016/j.iswcr.2015.05.002

[86] Kargas G, Kerkides P, Poulovassilis A. Infiltration of rain water in semi-arid areas under three land surface treatments. Soil and Tillage Research. 2012;**120**:15-24

[87] Pagliai M, Vignozzi N, Pellegrini S. Soil structure and the effect of management practices. Soil and Tillage Research. 2004;**79**:131-143

[88] Hei J, Xu J, Wang Y, Liu X, Yin X, Wang C, et al. Effects of Spartina alterniflora invasion on soil carbon, nitrogen, phosphorus, and CH_4, CO_2 emissions in Minjiang river estuarine wetland. Acta Scientiae Circumstantiae. 2022;**42**:416-426

[89] Bolinder M, Kätterer T, Andrén O, Ericson L, Parent L-E, Kirchmann H. Long-term soil organic carbon and nitrogen dynamics in forage-based crop rotations in northern Sweden (63-64N). Agriculture, Ecosystems and Environment. 2010;**138**(3):335-342

[90] Hati KM, Swarup A, Dwivedi A, Misra A, Bandyopadhyay K. Changes in soil physical properties and organic carbon status at the topsoil horizon of a vertisol of Central India after 28 years of continuous cropping, fertilization and manuring. Agriculture, Ecosystems and Environment. 2007;**119**(1):127-134

[91] Holland J. The environmental consequences of adopting conservation tillage in Europe: Reviewing the evidence. Agriculture, Ecosystems and Environment. 2004;**103**(1):1-25

[92] Davies DB, Finney JB. Reduced Cultivations for Cereals: Research, Development and Advisory Needs under Changing Economic Circumstances. Kenilworth: Home Grown Cereals Authority; 2002

[93] Barrios E. Soil biota, ecosystem services and land productivity. Ecological Economics. 2007;**64**:269-285. DOI: 10.1016/j.ecolecon

[94] Jiang X, Haddix ML, Cotrufo MF. Interactions between biochar and soil organic carbon decomposition: Effects of nitrogen and low molecular weight carbon compound addition. Soil Biology and Biochemistry. 2016;**100**:92-101. DOI: 10.1016/j.soilbio.2016.05.020

[95] CTIC, Conservation Tillage Information Center. National Crop Residue Management Survey. West Lafayette, In: Conservation Technology Information Center, Indiana. Available from: https://www.ctic.org/resource_display/?id=255&title=crop+residue+management+survey [Accessed: February 20, 2024]

[96] Kumar T, Kundu MS, Jha RK. Impact of crop rotation and tillage operations on mitigating greenhouse gas emissions and evaluation of sustainability index in rice- wheat-green gram cropping system of North Bihar. Journal of Environmental Management. 2024;**366**:121689. DOI: 10.1016/j.jenvman.2024.121689

[97] Mielke LN, Wilhelm WW. Comparisons of soil physical

characteristics in long-term tillage winter wheat fallow tillage experiments. Soil and Tillage Research. 1998;**49**(1-2):29-35

[98] Anikwe MAN, Mbah CN, Ezeaku PI, Onyia VN. Tillage and plastic mulch effects on soil properties and growth and yield of cocoyam (*Colocasia esculenta*) on an ultisol in southeastern Nigeria. Soil and Tillage Research. 2007;**93**:264-272

[99] Sarkar S, Singh SR. Interactive effect of tillage depth and mulch on soil temperature, productivity and water use pattern of rainfed barley (*Hordium vulgare* L.). Soil and Tillage Research. 2007;**92**:79-86

[100] Glab T, Kulig B. Effect of mulch and tillage system on soil porosity under wheat (*Triticum aestivum*). Soil and Tillage Research. 2008;**99**:169-178

[101] Franzluebbers AJ. Soil organic matter stratification ratio as an indicator of soil quality. Soil and Tillage Research. 2002;**66**:95-106

[102] Bhatt R, Khera K. Effect of tillage and mode of straw mulch application on soil erosion in the submountainous tract of Punjab, India. Soil and Tillage Research. 2006;**88**:107-115

[103] Puget P, Lal R. Soil organic carbon and nitrogen in Mollisol in Central Ohio as affected by tillage and land use. Soil and Tillage Research. 2005;**80**:201-213

[104] Zhang H. Organic matter incorporation affects mechanical properties of soil aggregates. Soil and Tillage Research. 1994;**31**:263-275

[105] Singh RC, Lenka S, Singh CD. Conservation tillage and manure effect on soil aggregation, yield and energy requirement for wheat in vertisols. Indian Journal of Agricultural Sciences. 2014;**84**(2):267-271

[106] Pearsons KA, Omondi EC, Zinati G, Smith A, Rui Y. A tale of two systems: Does reducing tillage affect soil health differently in long-term, side-by-side conventional and organic agricultural systems? Soil and Tillage Research. 2023;**226**:105562. DOI: 10.1016/j.still.2022.105562

[107] Vlček L, Šípek V, Zelíková N, Čáp P, Kincl D, Vopravil J. Water retention and infiltration affected by conventional and conservational tillage on a maize plot; rainfall simulator and infiltrometer comparison study. Agricultural Water Management. 2022;**271**:107800. DOI: 10.1016/j.agwat.2022.107800

[108] Krauss M, Wiesmeier M, Don A, Cuperus F, Gattinger A, Gruber S, et al. Reduced tillage in organic farming affects soil organic carbon stocks in temperate Europe. Soil and Tillage Research. 2022;**216**:105262. DOI: 10.1016/j.still.2021.105262

[109] Šarauskis E, Vaitauskienė K, Romaneckas K, Jasinskas A, Butkus V, Kriaučiūnienė Z. Fuel consumption and CO_2 emission analysis in different strip tillage scenarios. Energy. 2017;**118**:957-968. DOI: 10.1016/j.energy.2016.10.121

[110] De Almeida WS, Panachuki E, De Oliveira PTS, Da Silva MR, Sobrinho TA, De Carvalho DF. Effect of soil tillage and vegetal cover on soil water infiltration. Soil and Tillage Research. 2017;**175**:130-138. DOI: 10.1016/j.still.2017.07.009

[111] Liebhard G, Klik A, Neugschwandtner RW, Nolz R. Effects of tillage systems on soil water distribution, crop development, and evaporation and transpiration rates of soybean. Agricultural Water Management.

2022;**269**:107719. DOI: 10.1016/j.agwat.2022.107719

[112] Liu X, Song X, Li S, Liang G, Wu X. Understanding how conservation tillage promotes soil carbon accumulation: Insights into extracellular enzyme activities and carbon flows between aggregate fractions. Science of the Total Environment. 2023;**897**:165408. DOI: 10.1016/j.scitotenv.2023.165408

[113] Li Z, Zhang Q, Li Z, Qiao Y, Du K, Tian C, et al. Effects of no-tillage on greenhouse gas emissions in maize fields in a semi-humid temperate climate region. Environmental Pollution. 2022;**309**:119747. DOI: 10.1016/j.envpol.2022.119747

[114] Singh K, Singh M, Mishra SK, Soufan W, El Sabagh A. Reduced tillage and subsurface fertigation improve productivity and economic benefits in the cotton-wheat cropping system. Frontiers in Sustainable Food Systems. 2023;7:1185805. DOI: 10.3389/fsufs.2023.1185805

[115] Thapa B, Dura R. A review on tillage system and no-till agriculture and its impact on soil health. Archives of Agriculture and Environmental Science. 2024;**9**:612-617. DOI: 10.26832/24566632.2024.0903028

[116] Thomas G, Dalal R, Standley J. No-till effects on organic matter, pH, cation exchange capacity and nutrient distribution in a Luvisol in the semi-arid subtropics. Soil and Tillage Research. 2007;**94**(2):295-304. DOI: 10.1016/j.still.2006.08.005

[117] Che Y, Rejesus RM, Cavigelli MA, White KE, Aglasan S, Knight LG, et al. Long-term economic impacts of no-till adoption. Soil Security. 2023;**13**:100103. DOI: 10.1016/j.soisec.2023.100103

[118] Skaalsveen K, Ingram J, Clarke LE. The effect of no-till farming on the soil functions of water purification and retention in North-Western Europe: A literature review. Soil and Tillage Research. 2019;**189**:98-109. DOI: 10.1016/j.still.2019.01.004

[119] Blanco-Canqui H, Ruis SJ. No-tillage and soil physical environment. Geoderma. 2018;**326**:164-200. DOI: 10.1016/j.geoderma.2018.03.011

[120] Farahani S, Fard F, Asoodar M. Effects of contour farming on runoff and soil erosion reduction: A review study. Elixir Agriculture. 2016;**101**:44089-44093

[121] Jia L, Zhao W, Zhai R, An Y, Pereira P. Quantifying the effects of contour tillage in controlling water erosion in China: A meta-analysis. Catena. 2020;**195**:104829. DOI: 10.1016/j.catena.2020.10482908:50 22.02.2025

[122] Saggau P, Kuhwald M, Duttmann R. Effects of contour farming and tillage practices on soil erosion processes in a hummocky watershed. A model-based case study highlighting the role of tramline tracks. Catena. 2023;**228**:107126. DOI: 10.1016/j.catena.2023.107126

[123] Chen X, Liang Z, Zhang Z, Zhang L. Effects of soil and water conservation measures on runoff and sediment yield in red soil slope farmland under natural rainfall. Sustainability. 2020;**12**(8):3417. DOI: 10.3390/su12083417

[124] Gathagu JN, Mourad KA, Sang J. Effectiveness of contour farming and filter strips on ecosystem services. Water. 2018;**10**(10):1312. DOI: 10.3390/w10101312

[125] Chaudhary C, Yadav DB, Hooda VS, Chaudhary A, Parshad J, Kumar A, et al. Rice residue management alternatives

and nitrogen optimization: Impact on wheat productivity, microbial dynamics, and enzymatic activities. Frontiers in Sustainable Food Systems. 2024;**8**:1402803. DOI: 10.3389/fsufs.2024.1402803

[126] Tian L, Zhang S. Effects of strip cropping with reducing row spacing and super absorbent polymer on yield and water productivity of oat (*Avena sativa* L.) under drip irrigation in Inner Mongolia, China. Scientific Reports. 2022;**12**:11441. DOI: 10.1038/s41598-022-15418-w

[127] Głowacka A. The effects of strip cropping and weed control methods on yield and yield components of dent maize, common bean and spring barley. Polish Journal of Natural Science. 2013;**28**:389-408

[128] Campanelli G, Iocola I, Leteo F, Montemurro F, Platani C, Testani E, et al. Strip cropping in organically managed vegetable systems: Agronomic and environmental effects. Renewable Agriculture and Food Systems. 2023;**38**:e31. DOI: 10.1017/S1742170523000248

[129] Mcmillen M. The Effect of Mulch Type and Thickness on the Soil Surface Evaporation Rate. USA: California Polytechnic State University; 2013

[130] Ye L, Xu Y, Zhu G, Zhang W, Jiao Y. Effects of different mulch types on farmland soil moisture in an artificial oasis area. Land. 2023;**13**(1):34. DOI: 10.3390/land13010034

[131] Khamare Y, Marble SC. Mulching as a weed management tool in container plant production - review. Frontiers in Agronomy. 2023;**5**:1235196. DOI: 10.3389/fagro.2023.1235196

[132] Douibi M, Krishtammagari A, Sánchez-Martín MJ, Rodríguez-Cruz MS, Marín-Benito JM. Mulching vs. organic soil amendment: Effects on adsorption-desorption of herbicides. Science of the Total Environment. 2023;**892**:164749. DOI: 10.1016/j.scitotenv.2023.164749

[133] Modiba MM, Ocansey CM, Ibrahim HTM, Birkás M, Dekemati I, Simon B. Assessing the impact of tillage methods on soil moisture content and crop yield in Hungary. Agronomy. 2024;**14**(8):1606. DOI: 10.3390/agronomy14081606

[134] Chalise D, Kumar L, Sharma R, Kristiansen P. Assessing the impacts of tillage and mulch on soil erosion and corn yield. Agronomy. 2020;**10**(1):63. DOI: 10.3390/agronomy10010063

[135] Toth M, Stumpp C, Tıkla A, Strauss P, Mehdi B, Liebhard G, et al. Long-term effects of tillage systems on soil health of a silt loam in lower Austria. Soil and Tillage Research. 2024;**241**:106120. DOI: 10.1016/j.still.2024.106120

[136] Liu Z, Cao S, Sun Z, Wang H, Qu S, Lei N, et al. Tillage effects on soil properties and crop yield after land reclamation. Scientific Reports. 2021;**11**(1):1-12. DOI: 10.1038/s41598-021-84191-z

[137] Steponavičienė V, Rudinskienė A, Žiūraitis G, Bogužas V. The impact of tillage and crop residue incorporation systems on agrophysical soil properties. Plants. 2023;**12**(19):3386. DOI: 10.3390/plants12193386

[138] Fawzi H, Mostafa SA, Ahmed D, Alduais N, Mohammed MA, Elhoseny M. TOQO: A new tillage operations quality optimization model based on parallel and dynamic decision support system. Journal of Cleaner Production. 2021;**316**:128263. DOI: 10.1016/j.jclepro.2021.128263

[139] Li J, Yk W, Guo Z. Effects of conservation tillage on soil physicochemical properties and crop yield in an arid loess plateau, China. Scientific Reports. 2020;**10**:4716. DOI: 10.1038/s41598-020-61650-7

[140] Yağanoğlu E, Aydın A. The effect of organic and chemical fertilizers on the activity of soil enzymes in soils of different compositions. Turkish Journal of Agriculture and Forestry. 2024;**48**(1):11. DOI: 10.55730/1300-011X.3167

[141] Harju AV, Närhi I, Mattsson M, Kerminen K, Kontro MH. Organic matter causes chemical pollutant dissipation along with adsorption and microbial degradation. Frontiers in Environmental Science. 2021;**9**:666222. DOI: 10.3389/fenvs.2021.666222

[142] Liu Q, Cui H, Yang W, Wang F, Liao H, Zhu Q, et al. Soil conditioner improves soil properties, regulates microbial communities, and increases yield and quality of *Uncaria rhynchophylla*. Scientific Reports. 2024;**14**(1):1-11. DOI: 10.1038/s41598-024-64362-4

[143] Enesi RO, Dyck M, Chang S, Thilakarathna MS, Fan X, Strelkov S, et al. Liming remediates soil acidity and improves crop yield and profitability - a meta-analysis. Frontiers in Agronomy. 2023;**5**:1194896. DOI: 10.3389/fagro.2023.1194896

[144] Ehtaiwesh AF. The effect of salinity on nutrient availability and uptake in crop plants. 2022;**9**:55-73. DOI: 10.47891/sabujas.v0i0.55-73

[145] Mandal D, Patra S, Sharma NK, Alam NM, Jana C, Lal R. Impacts of soil erosion on soil quality and agricultural sustainability in the North-Western Himalayan region of India. Sustainability. 2023;**15**(6):5430. DOI: 10.3390/su15065430

[146] Faraj TK, EL-Saeid MH, Najim MMM, Chieb M. The impact of pesticide residues on soil health for sustainable vegetable production in arid areas. Separations. 2024;**11**(2):46. DOI: 10.3390/separations11020046

[147] Conradi Elio J, Gonçalves A Jr, Seidel E, Lindner Z, Zimmermann J, de Dias O, et al. Effects of liming on soil physical attributes: A review. Journal of Agricultural Science. 2024;**12**:278-278. DOI: 10.5539/jas.v12n10p278

[148] Makanda K, Nzama S, Kanyerere T. Assessing the role of water resources protection practice for sustainable water resources management: A review. Water. 2021;**14**(19):3153. DOI: 10.3390/w14193153

[149] Kharlukhi DG, Upadhyaya K, Sahoo UK. Influence of integrated nutrient management on soil health, growth and yield of paddy in "jhum lands" of North-Eastern Himalayas. Discover Agriculture. 2024;**2**:107. DOI: 10.1007/s44279-024-00128-w

[150] Gamage A, Gangahagedara R, Gamage J, Jayasinghe N, Kodikara Suraweera NP, Merah O. Role of organic farming for achieving sustainability in agriculture. Farming System. 2023;**1**(1):100005. DOI: 10.1016/j.farsys.2023.100005

[151] Bwalya B, Mutandwa E, Chiluba BC. Awareness and use of sustainable land management practices in smallholder farming systems. Sustainability. 2023;**15**(20):14660. DOI: 10.3390/su152014660

[152] Tiong YW, Sharma P, Xu S, Bu J, Ve S, Foo ÜBL, et al. Enhancing sustainable crop cultivation: The impact of renewable soil amendments and digestate fertilizer on crop growth and nutrient composition. Environmental

Pollution. 2024;**342**:123132. DOI: 10.1016/j.envpol.2023.123132

[153] Dewangan S, Kumar SS, Kumari L, Minj P, Kumari J, Sahu R. The effects of soil ph on soil health and environmental sustainability: A review. Journal of Emerging Technologies and Innnovative Research. 2023;**10**:611-616

[154] Oyetunji O, Bolan N, Hancock G. A comprehensive review on enhancing nutrient use efficiency and productivity of broadacre (arable) crops with the combined utilization of compost and fertilizers. Journal of Environmental Management. 2022;**317**:115395. DOI: 10.1016/j.jenvman.2022.115395

[155] Shanmugavel D, Rusyn I, Solorza-Feria O, Kamaraj S. Sustainable SMART fertilizers in agriculture systems: A review on fundamentals to in-field applications. Science of the Total Environment. 2023;**904**:166729. DOI: 10.1016/j.scitotenv.2023.166729

[156] Blanchy G, Bragato G, Di BC, Jarvis N, Larsbo M, Meurer K, et al. Soil and crop management practices and the water regulation functions of soils: A synthesis of meta-analyses relevant to European agriculture. 2022. DOI: 10.5194/egusphere-2022-270

[157] Zhou W, Li C, Wang S, Ren Z, Stringer LC. Effects of vegetation restoration on soil properties and vegetation attributes in the arid and semi-arid regions of China. Journal of Environmental Management. 2023;**343**:118186. DOI: 10.1016/j.jenvman.2023.118186

[158] Candel-Pérez D, Lucas-Borja ME, Plaza-Álvarez PA, Carmona YMD, Soria R, Ortega R, et al. Effects of grazing on soil properties in Mediterranean forests (central-eastern Spain). Journal of Environmental Management. 2024;**354**:120316. DOI: 10.1016/j.jenvman.2024.120316

[159] Silveira ML, Kohmann MM. Maintaining soil fertility and health for sustainable pastures. Management Strategies for Sustainable Cattle Production in Southern Pastures. Florida, United States: Academic Press; 2019:35-58. DOI: 10.1016/B978-0-12-814474-9.00003-7

[160] Hamidov A, Helming K, Bellocchi G, Bojar W, Dalgaard T, Ghaley BB, et al. Impacts of climate change adaptation options on soil functions: A review of European case-studies. Land Degradation & Development. 2018;**29**(8):2378. DOI: 10.1002/ldr.3006

[161] Bach EM, Ramirez KS, Fraser TD, Wall DH. Soil biodiversity integrates solutions for a sustainable future. Sustainability. 2020;**12**(7):2662. DOI: 10.3390/su12072662

[162] Kuzucu M, Gul I. Erosion Control in Pastures (in Turkish). In book: Strategies for Reforestation and Sustainable Fruit Farming. 2024. pp. 45-72. DOI: 10.69860/nobel.9786053359395.3

[163] Huang X, Lu Q, Yang F. The effects of farmers' adoption behavior of soil and water conservation measures on agricultural output. International Journal of Climate Change Strategies and Management. 2020;**2**(5):599-615. DOI: 10.1108/IJCCSM-02-2020-0014

[164] Mao W, Zhu Y, Wu J, Ye M, Yang J. Evaluation of effects of limited irrigation on regional-scale water movement and salt accumulation in arid agricultural areas. Agricultural Water Management. 2022;**262**:107398. DOI: 10.1016/j.agwat.2021.107398

[165] Singh A. Effective management of water resources problems in irrigated agriculture through simulation modeling. Water Resources Management: An International Journal. 2024;**38**(8):2869-2887

[166] Cordel J, Prämaßing W, Anlauf R. Impact of rootzone construction and irrigation methods on soil moisture in sports fields under greenhouse conditions. European Journal of Horticultural Science. 2024;**89**:1-14. DOI: 10.17660/eJHS.2024/007

[167] Jiao F, Ding R, Du T, Kang J, Tong L, Gao J, et al. Multi-growth stage regulated deficit irrigation improves maize water productivity in an arid region of China. Agricultural Water Management. 2024;**297**:108827. DOI: 10.1016/j.agwat.2024.108827

[168] Dwijendra NKA, Salih S. The effect of various irrigation technologies and strategies on water resources management. Journal of Water and Land Development. 2022:143-147. DOI: 10.24425/jwld.2022.140790

[169] Zahra N, Farooq M, Siddique K. Strategies for improving water use efficiency in dryland agroecosystems. Frontiers of Agricultural Science and Engineering. 2021;**8**:2021409. DOI: 10.15302/J-FASE-2021409

[170] Alnaimy MA, Shahin SA, Vranayova Z, Zelenakova M, Abdel-Hamed EMW. Long-term impact of wastewater irrigation on soil pollution and degradation: A case study from Egypt. Water. 2021;**13**(16):2245. DOI: 10.3390/w13162245

[171] Laskookalayeh S, Mardani M, Ali S. Investigating the effects of management of irrigation water distribution on farmers' gross profit under uncertainty: A new positive mathematical programming model. Journal of Cleaner Production. 2022;**351**:131277. DOI: 10.1016/j.jclepro.2022.131277

[172] Pashaev E. Eco-meliorative monitoring and evaluation of the state of irrigated lands in the influence zone of the system of Samur-Absheron channel in connection with its reconstruction. Russian Agricultural Sciences. 2016;**42**:160-162. DOI: 10.3103/S1068367416020117

[173] Hamidov A, Kasymov U, Djumaboev K, Paul C. Rebound effects in irrigated agriculture in Uzbekistan: A stakeholder-based assessment. Sustainability. 2022;**14**(14):8375. DOI: 10.3390/su14148375

[174] Hatfield JL. Environmental impact of water use in agriculture. Agronomy Journal. 2022;**107**(4):1554-1556. DOI: 10.2134/agronj14.0064

[175] Jahangirpour D, Zibaei M. Farmers' decision to adoption of modern irrigation systems under risk condition: Application of stochastic efficiency with respect to a function approach. Frontiers in Water. 2022;**4**:931694. DOI: 10.3389/frwa.2022.931694

[176] Zhang Z, Yao W, Huang Y, Jiang X, Gao Z, Chen S, et al. Irrigation water salinity affects solute transport and its potential factors influencing salt distribution in unsaturated homogenous red soil. Agronomy. 2024;**14**(11):2453. DOI: 10.3390/agronomy14112453

[177] Mishra S, Kumar R, Kumar M. Use of treated sewage or wastewater as an irrigation water for agricultural purposes- environmental, health, and economic impacts. Total Environment Research Themes. 2023;**6**:100051. DOI: 10.1016/j.totert.2023.100051